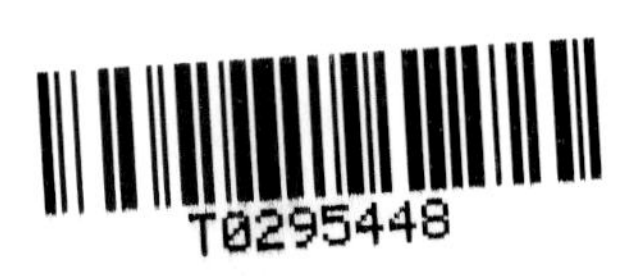
T0295448

Quantitative Phase Field Modelling of Solidification

Quantitative Phase Field Modelling of Solidification

Nikolas Provatas
Tatu Pinomaa
Nana Ofori-Opoku

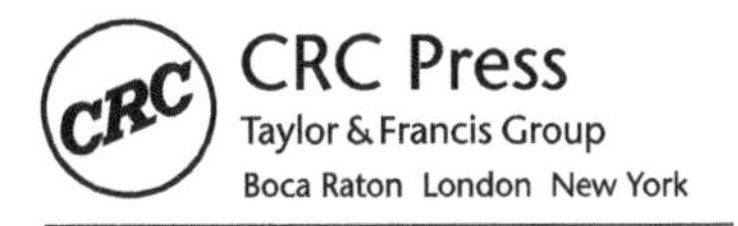

CRC Press is an imprint of the
Taylor & Francis Group, an **informa** business

First edition published 2022
by CRC Press
6000 Broken Sound Parkway NW, Suite 300, Boca Raton, FL 33487-2742

and by CRC Press
2 Park Square, Milton Park, Abingdon, Oxon, OX14 4RN

CRC Press is an imprint of Taylor & Francis Group, LLC

Nana Ofori-Opoku prepared this publication in his personal capacity. The opinions and material expressed in this publication are the author's own and do not reflect the view of Canadian Nuclear Laboratories Ltd., Atomic Energy of Canada Limited, or the Government of Canada

ISBN: 978-0-367-76857-7 (hbk)
ISBN: 978-1-032-06888-6 (pbk)
ISBN: 978-1-003-20431-2 (ebk)

DOI: 10.1201/9781003204312

Typeset in Latin Modern font
by KnowledgeWorks Global Ltd.

Contents

SECTION II

SECTION V

Authors

Nikolas Provatas is a Professor of physics at McGill University and holds a Canada Research Chair (Tier 1) in Computational Materials Science. He is also the Scientific Director of the McGill High Performance Computing Centre. From 2001 to 2012, he was a Professor of Materials Science and Engineering at McMaster University. His research uses high-performance computing, dynamic adaptive mesh refinement techniques, condensed matter physics, and experimentation to understand the fundamental origins of nano-microstructure pattern formation in non-equilibrium phase transformations, and the role of microstructures in materials processes. He has numerous scientific contributions in the understanding of length scale selection in dendritic solidification and meta-stable phase formation in solid-state transformations in metal alloys.

Tatu Pinomaa is a Senior Scientist at VTT Technical Research Centre of Finland Ltd. He received a doctor of science (tech) degree from Aalto University (Finland), where he developed phase field modelling techniques to investigate rapid solidification microstructures in metal additive manufacturing conditions. His current research combines various computational approaches to predict the formation, evolution, and micromechanical response of metallic microstructures for industrial applications.

Nana Ofori-Opoku is a Research Scientist at Canadian Nuclear Laboratories Ltd. He received a doctorate in materials science from McMaster University, where he explored computational models for microstructure evolution in materials. He did his postdoctoral work at McGill University, followed by a NIST-CHiMaD fellowship at Northwestern University and the National Institute of Standards and Technology. His research continues to develop theoretical and computational tools to study microstructure evolution in nuclear materials and the dynamics of phase transformations.

CHAPTER 1

A Brief History of Phase Field Modelling

IN THE PAST TWO TO THREE decades or so, phase field models have seen an explosive growth in their use for exploring non-equilibrium phase transformations and microstructure evolution in materials science and engineering. They are appealing for describing phase transformations since they can be connected with statistical thermodynamics once an appropriate set of order parameters and their symmetries are identified for a particular system. Dynamical equations of motion for one or more order parameter fields can also be formulated based on dissipative free energy minimization using minimization principles and conservation laws. Phase field models are very simple to code and require no overhead in terms of tracking moving, impinging and coalescing interfaces, which is the quintessential feature of non-equilibrium microstructure processes. Phase field models have been applied to a wide range of phenomena ranging from single-crystal solidification, directional and polycrystalline solidification, grain growth, elasticity and precipitation. These days they can also be used quantitatively for select phenomena in solidification and solid state transformations, making them viable tools for materials science and engineering. Quantitative calculations with phase field models requires detailed asymptotic boundary layer analysis, which prescribes how to match their behaviour with appropriate sharp interface models in the limit when a suitable scale separation exists between the length of diffusion of impurities (or heat) and phase coexistence boundaries (hereafter called phase field interfaces). The remainder of this chapters goes through a brief history of some of the major milestones in phase field

DOI: 10.1201/9781003204312-1

modelling in recent years. Of course, the complete story is far too long to tell in a Concise Book series text like this, whose main focus is *not* a history lesson.

One of the first phase field models of solidification of a pure substance was by Colins and Levine [1], which used the "model C" of the Hohenberg and Halperin classification [2] to demonstrate thermally-controlled solidification. Karma and Rappel [3] later showed how using matched asymptotic boundary layer analysis can map a phase field model of a pure material quantitatively onto the dynamics of the corresponding Stefan model (hereafter *sharp-interface model*) of solidification of a pure material [4] while operating with *thin* -rather than atomically sharp-order parameter interfaces. Specifically, it decoupled the time scale of phase field simulations from the atomic kinetics through the interface, a feature absent from so-called first order asymptotic mapping of the phase field model onto the sharp-interface approach originally derived by Caginalp [5]. At around the same time, Provatas et al. developed a novel dynamic adaptive mesh refinement algorithm [6,7] for phase field models that dramatically reduced the computational barrier of grid discretization that makes data management and CPU times associated with large scale and long-time simulations of free boundary problems intractable.

In the scope of alloy materials, the work of Warren and Boettinger developed a specialized two-phase alloy model that demonstrated many of the qualitative features of casting could be modelled by coupling an order parameter to a solute concentration field [8]. Karma and co-workers [9, 10] adapted the thin-interface asymptotic analysis for a two-phase field model of solidification of an ideal binary alloy. Attaining the sharp interface limit for alloy solidification is made difficult by the large disparity in solute diffusion between the solid and liquid. Specifically, spurious solute trapping and lateral diffusion effects are generated through the interface. While physical in origin, these effects are negligible at low rates of solidification since true interface widths in metals are on the scale of a few *nm*. However, when employing diffuse interfaces for numerical expediency, these spurious corrections become undesirably exaggerated. To remedy this, Refs. [9, 10] introduced a so-called *anti-trapping flux* to correct for spurious solute trapping caused by diffuse numerical interfaces. They also employed specific interpolation functions in the driving force of the free energy, which effectively transforms the driving force for solidification into the difference in grand potential between phases[1],

[1]Although this was not made explicit in these works originally.

which leads to complete decoupling of the solute and order parameter fields in equilibrium. This makes it possible to (i) select an arbitrarily diffuse phase field interface width and (ii) tune the surface energy of a phase field interface to any desired value independently of the solute field, a critical feature for quantitative modelling of solidification. Ramirez et al. [11] later adapted the matched asymptotic analysis to a single-phase alloy model coupled to thermal transport. Their results confirmed that the thin interface asymptotic analysis of Ref. [10] essentially continue to hold for low cooling rates even for non-isothermal conditions.

Subsequent work by Tong et al. [12] adapted the thin interface asymptotic analysis of Ref. [10] for the more general case of two-phase solidification in non-isothermal binary alloys. At the heart of their approach was approximating the free energy of the liquid or any solid near equilibrium using the ideal binary alloy free energy as a mathematical *fitting function* with temperature-dependent parameters. Ofori-Opoku et al. [13] later generalized the interpolation functions used in Ref. [10] into a vectorial form in order to interpolate between liquid and *multiple* crystal grains (i.e. orientations), each represented by its own order parameter. For the case of ideal alloys, this approach is effectively a *multi-order parameter* variant of the model in Ref. [10]. By utilizing the approach of Tong et al. [12] to fit the liquid or solid free energy close to their equilibrium concentrations, the model of Ref. [13] effectively provides a quantitative *multi-order parameter phase field model* for simulating poly-crystalline, non-isothermal solidification of *non-ideal* alloys.

Phase field models with multiple order parameters go back to the original multiple-order parameter models introduced by Khachaturyan [14] for the study of ordered precipitates and other solid state phase transformations. The models consider the phase field as a physical order parameter that distinguishes between an ordered and disordered phase. The interface free energy is defined by gradients in the order parameters and interactions between ordered phases through a polynomial expansion in the corresponding order parameters[2]. Their extension to multi-component alloys couples a solute-dependent chemical energy contribution to the free energy that is interpolated between different phases by the variation of the order parameter fields [19–21]. For quantitative control of interface energies, decoupling of equilibrium interface

[2]These model bear some resemblance to the scalar order parameter limit of complex amplitude equations that come out of coarse graining of so-called phase field crystal (*PFC*) models [15–18]

energy from the solute fields (as discussed above in the context of quantitative modelling of solidification) is achieved in this approach by introducing fictitious, or *auxiliary*, concentration fields corresponding to each phase, as in Refs. [22–24] (more on this method below). Dynamics in multi-order parameter models follow the usual dissipative free energy minimization for each order parameter and mass conservation dynamics for solute(s), both coupled to constraints imposed by the auxiliary concentration fields. The multi-order parameter approach has been adopted by several groups to model grain growth in solid state systems [25, 26] and multi-component alloys [13, 20, 21, 27–29].

Another, perhaps more physical, line of research for modelling polycrystalline solidification involves the use of an order parameter that self-consistently allows for solid-liquid ordering *and* orientation control. The first works in this direction go back to Morin and Grant [30], which created a multi-well free energy based on a vector order parameter that had minima at multiple orientations. While not rotationally invariant, it was the first step toward assigning multiple angles needed to emulate polycrystalline orientations using a single order parameter field. A different approach was later developed by the works of Kobayashi and Warren [31–33] that coupled a scalar order parameter field to a rotationally invariant orientational field. This approach allows an arbitrary number of orientations to be simulated. Orientational phase field methods were first developed for pure materials and then extended to binary alloys by Granasy and co-workers [34]. When coupled with stochastic fluctuations, the orientational field approach offers one of the most physically consistent approaches for simulating nucleation in polycrystalline solidification. Besides the fundamental and self-consistent nature of this approach, it is also quite practical as it allows a significant reduction on computational overhead to model multiple orientations. Another benefit of the orientational field approach is that nucleation of multiple phases via noise is done simultaneously through the orientational fields, thus avoiding inconsistencies with nucleating multiple phases in multi-phase field approaches via noise. By way of contrast, multi-order parameter models (or multi-phase field models, discussed below) can only handle on the order of 5–10 orientations and phases before they become computationally inefficient on a small number of processors. Moreover, such models cannot handle noise-induced nucleation since it can lead to problems with overlapping of different phase fractions. Of course, all models have drawbacks, and one drawback of such orientation order parameter models is that the free energy is non-analytic, making them difficult to rationalize from basic principles[3]. Another drawback is that there is presently no asymptotic analysis available through which to map orientational order parameter models onto an appropriate sharp interface model.

A separate avenue for modelling multiple phases in multi-component alloys is the so-coined *multi-phase field models*, developed by Steinbach and collaborators [41–47], which is applicable to both solidification and solid state transformations. The phenomenology of these models assigns each phase a *phase-fraction* field[4] which are constrained to sum to one. The free energy is interpolated across interfaces or n-junctions by a phase-fraction-weighted superposition of each phase's free energy[5]; as with some of the multi-order parameter models discussed above, the free energy of each phase is written as a function of "auxiliary concentrations" assigned to that phase. Interactions between phase fraction fields are implemented via a so-called double obstacle potential, or other polynomial interaction terms between the phase fractions. Interface energy is modelled by various phenomenological gradient operator forms acting on the phase fractions. The evolution of the phase fraction fields follows non-conserved dynamics subject to the constraint that the phase fractions sum to one, through the use of Lagrange multipliers[6]. Solute fields are updated using the usual mass transport equations, coupled to constraints imposed by the auxiliary concentration fields.

The *auxiliary* concentration field method used in multi-phase field models (and some multi-order parameter models) is crucial for decoupling solute and phase fields—a useful ingredient for quantitative modelling of surface energy—and merits a little deeper explanation as it will be shown to naturally emerge out of the approach adopted in this

[3]Alternate attempts to self-consistently couple orientation and ordering employ phase field models with complex order parameters derived by coarse graining *phase field crystal (PFC)* theories of pure materials [35] and alloys [36–38], or classical density functional theories [39]. While analytically consistent such models also suffer from orientation issues at large angles [40]

[4]The interpretation of the phase filed is not practically important in the sharp interface limit of solidification, but is important when connecting phase field models arising from microscopic theories.

[5]Moelans [48] showed that phase-fraction weighting of the free energy can also be achieved in multi-order parameter alloy models by defining phase fraction fields in terms of traditional order parameters, that need not explicitly sum to one everywhere.

[6]Using Lagrange multipliers in the dynamics can cause unphysical non-local effects whereby unwanted phases appear in triple-junctions. This is addresed by using higher order polynomial interactions between phase fractions to penalize such effects.

book. It was developed by Kim, Kim and Suzuki [23, 24] for a two-phase binary alloy and Tiaden and co-workers [22] for multi-phase and multi-component alloys. We explain this decoupling approach here for the simple case of a two-phase binary alloy. The approach of Tiaden et al. [22] defines a *phase fraction* field for each phase (e.g. one for solid, one for liquid). Physical solute is then defined as a phase fraction-weighted superposition of these fictitious (*auxiliary*) solute fields (one for each phase). The bulk free energy is similarly expanded as a phase fraction-weighted sum of each phase's free energy, where the free energy of each phase is expressed in terms of its auxiliary concentration. (Ref. [23] follows a similar approach, except the physical definition of the phase field as an order parameter between solid and liquid is retained.) The time evolution of the phase fractions follows the usual free energy minimization, constrained in the case of phase fractions to sum to one through a Lagrange multiplier. Solute dynamics follow the usual mass conservation dynamics, wherein solute is driven by gradients in the chemical potential. The chemical potential, however, is written in terms of the auxiliary concentration fields, and so these must first be updated. The update of the auxiliary concentration fields is done by applying the constraint that the chemical potentials corresponding to each auxiliary concentration must be equal within each volume element, at each time step. This effectively transforms the driving force in the phase field equation to the difference in local grand potential densities between the two phases (see Ref. [38] for detailed derivation). While mathematically laborious, this approach allows multi-component models to decouple the equilibrium interface energy from the solute profile through the interface, for any interface width. In effect, this methods trades the freedom afforded by the choice of interpolation functions used in Refs. [10, 13] by the use of auxiliary concentration fields.

Despite the important benefit of decoupling solute and order parameter fields discussed above, multi-phase and multi-order parameter models are not immune from solute trapping and spurious interface kinetic effects discussed earlier. These are exacerbated with the use of diffuse phase field interfaces—which all phase field models must use for practical modelling of solidification. This phenomenon can typically be neglected to lowest order in solid-state transformations, where the rate of interface motion and the disparity of diffusion between phases is small. However, failure to capture the correct kinetics during the early stages of *solidification* can lead to incorrect interface structure and impurity micro-segregation in the final solidified system. Most of the early multi-phase field-type models described above suffered from this deficiency as they only employed so-called first order asymptotic interface analysis to match model parameters to the sharp interface limit [42]. An exception to this is the work of Folch and Plapp [49], which adapted the thin interface asymptotics of Ref. [10] (accurate to second-order in the interface width) for a multi-phase field model for eutectic solidification that approximated the free energy of each phase with a quadratic function in concentration centred at the minima of each phase's free energy. Another exception was the multi-order parameter binary alloy model of Ofori-Opoku and co-workers [13], which used a multi-dimensional extension of the interpolation functions of Ref. [10] for multiple order parameters (and added anti-trapping currents for each order parameter) to quantitatively emulate the appropriate sharp interface kinetics across any diffuse solid-liquid interface. Nowadays, most multi-phase field-type models utilize anti-trapping fluxes in mass conservation equations [50] to reduce spurious effects caused by diffuse interfaces and thus approximate local interface equilibrium as required to emulate the sharp interface limit.

Recently, Plapp [51] introduced an elegant re-formulation of the traditional two-phase binary alloy phase field model. This approach starts from the grand potential ensemble rather than the usual Helmholtz free energy approach. In this approach, the phase field retains its classic meaning as order parameters. The driving force for solidification becomes the *grand potential* difference between the solid and liquid, and the evolution of the *physical* concentration, an extensive variable, is replaced in favour of its conjugate field, the chemical potential, which is an intensive and natural variable of the grand potential ensemble. For the case of a constant diffusion coefficient, the description of two-phase binary alloy solidification becomes mathematically identical to the solidification of a pure material [4]; this makes it possible to re-cycle much of the asymptotic analysis machinery developed for a pure material [3] to parameterize the sharp interface limit of a two-phase binary alloy. Also, in this ensemble, the solute field, which is derivable from the grand potential, provides a natural interpolation of the "auxiliary" solute fields (one per phase). These are each derived from the grand potential density of that phase. Thus, the "grand potential phase field" approach naturally identifies a physical interpretation of the auxiliary fields previously introduced phenomenologically in works such as Refs. [22–24]. Moreover, by evolving the chemical potential directly, phase field models formulated in the grand potential ensemble do not require the computational cost of having to locally match chemical potentials defined by fictitious solute fields. Recently, Nestler and co-workers adapted the grand potential approach to the multi-phase field (phase fraction) approach to simulate solidification in multi-component alloys [50, 52].

Section I of this book shows how to use the grand potential ensemble to reformulate the multi-order-parameter phase field model of solidification of Ofori-Opoku et al. [13] into an easy-to-use multi-order parameter theory for multi-component solidification. This is done by deriving the equations of motion for multiple order parameters and multiple chemical potentials from a unified grand potential functional. By working in the grand potential ensemble, there is no need for specialized interpolation functions to decouple concentration and order parameters, the extension to multiple components becomes straightforward and the interpretation of all fields is at all times physical. Following the model's derivation, we examine its equilibrium properties in bulk phases and at interfaces, and show how to incorporate stochastic noise to simulate thermal fluctuations. In Section II of the book several specializations and applications of the model are examined. Section III discusses some general ideas about the interpretation of operating phase field model in the limit of a sharp interface model. Section IV of the book deals with non-equilibrium solidification theory, and shows how to manipulate the asymptotic analysis of the model derived in the first half of the book to emulate the so-called continuous growth model of solidification, a generally accepted model for non-equilibrium solidification at moderate solidification rates. Section V contains Appendices (A) and (B). Specific details of the remaining topics are broken down in detail in Chapter 2.

CHAPTER 2

Overview of This Book

THE REMAINDER OF THIS BOOK has four sections, summarized as follows:

- Section I (Chapters 3–8) begins by deriving a *multi-order parameter* model of solidification in the grand potential ensemble. We begin with a grand potential functional formulated in terms of multiple chemical potentials and phases. We derive the dynamical evolution equations for the order parameters and chemical potential fields, followed by an examination of the model's equilibrium properties, which include methods for computing the equilibrium concentration fields and the methods for calculating excess energies of solid-liquid and solid-solid interfaces. This section also includes methods for making phase field simulations of solidification for quantitative applications through the addition of anti-trapping fluxes to the chemical potential equations in order to control the level of solute trapping and other spurious effects caused by diffuse interfaces. We end this section by showing how to incorporate thermal fluctuations into the theory in order to allow simulation of side-branching and nucleation. The theories developed in Section I collectively serve as easy-to-use platforms for quantitatively modelling multi-component and multi-phase solidification.
- Section II (Chapter 9) specializes the generalized multi-order parameter grand potential phase field theory in Section I for three special applications: (A) deals with the case of single-phase, polycrystalline *multi-component* solidification in the so-called *local supersaturation* limit of each chemical component. This effectively extends the model of Ref. [13] to multiple components; (B)

DOI: 10.1201/9781003204312-2

examines *multi-phase* binary alloys whose free energies can be approximated by a 3-parameter quadratic form; (C) generalizes part (B) to multiple components by using a multi-dimensional quadratic form for the multi-component free energy of a phase. By formulating the equations of motion in terms of order parameters and supersaturation fields, the model in (C) is shown to reduce, at low undercooling, to a matrix analogue of the binary model in part (A), which is itself a multi-order parameter, multi-component version of the model in Ref. [10]. The models derived in sections (B) and (C) are also shown to contain higher order supersaturation terms in the driving force, which can be relevant at higher thermodynamic driving forces.

- Section III (Chapter 11) discusses the thin interface limit of phase field models. It starts by referring the reader to a second-order matched asymptotic analysis of a phase field model, derived fully in Appendix (B) Specifically, This appendix calculation shows how to extract the effective sharp interface kinetics of the solid-liquid interface in the case of a single-phase binary alloy, at low undercooling. The main text of Chapter 11 then goes on to discuss the parameter relations derived from Appendix (B) that map the dynamical phase field equations onto the classical sharp interface model of alloy solidification, focusing specifically on the interpretation of what it means to say that results from phase field simulations converge to the results this sharp interface model.

- Section IV (Chapter 13) begins by discussing the theory of the *continuous growth model of solidification* (CGM) pioneered analytically by Aziz and co-workers [53, 54]. CGM is a non-equilibrium sharp interface model that is often used for describing rapid solidification when the solid-liquid interface exhibits solute trapping and velocity-dependent corrections that cause the Gibbs-Thomson condition and local interface undercooling to deviate from the classic sharp interface model of solidification; this regime of solidification is becoming an increasingly important limit for advanced industrial metal processing techniques such as additive manufacturing. Following the theoretical introduction to CGM, the results of the asymptotic analysis in Chapter 11 are used to examine the detailed effects of the interface width on the solidification kinetics of a rapidly solidifying phase field interface. Here we specifically focus on showing how to asymptotically map the phase field model's thin interface behaviour onto the continuous growth model of solidification.

- Section V contains the following appendices: Appendix (A) couples the evolution of the phase field equations with heat transport. Appendix (B) contains the details of the asymptotic analysis of the main grand potential phase field equations derived and used in the main text of the book.

I

CHAPTER 3

Recap of Grand Potential Thermodynamics

WE BEGIN THIS CHAPTER WITH a quick review of the thermodynamics of the grand potential, which will be employed in this chapter to construct a phase field model, in contrast to the more traditional Helmholtz free energy (or entropy) typically employed in phase field modelling in the literature. The grand potential is defined by

$$\Omega = F - \sum_{i=1}^{n} \mu_i N_i, \tag{3.1}$$

where F is the Helmholtz free energy, μ_i is the chemical potential of species i and N_i is the number of particles for the ith species, and n denotes the total number of components in system. For each component, the chemical potential is defined by

$$\mu_i = \left.\frac{\partial F}{\partial N_i}\right|_{T,V,N_j \neq N_i}, \tag{3.2}$$

Equation (3.1) is a Legendre transformation of the Helmholtz free energy, from which it is straightforward to derive the well-known relation

$$d\Omega = -S\,dT - p\,dV - \sum_{i=1}^{n} N_i\,d\mu_i, \tag{3.3}$$

where T is the temperature, p is the pressure and S is the entropy of the system. Eq. (3.3) implies that the natural variables of Ω are T, V

DOI: 10.1201/9781003204312-3

and μ_i (whereas for the Helmholtz free energy they are S, V, N_i). Thus, by expressing Ω in terms of its natural variables, we can immediately calculate the number of particles N as

$$N_i = -\left.\frac{\partial \Omega}{\partial \mu_i}\right|_{T,V,\mu_j \neq \mu_i} \tag{3.4}$$

The description of alloys typically uses the concentration as a measure of the amount of each species, rather than particle number. To switch to concentration variables, some definitions are first required. We begin by noting that the sum of particle species satisfies $N_1 + N_2 + \cdots + N_n = N$, where N is the total number of particles in a system. As a result, we can write

$$\sum_{i=1}^{n} n_i = \frac{N}{V} = \bar{\rho} = \frac{N_A}{\nu_m}, \tag{3.5}$$

where $n_i = N_i/V$ is the number density of the ith species, $\bar{\rho}$ is the overal number density of the alloy, N_A is Avogadro's number and ν_m is the molar volume of the alloy. We can define the concentration of the ith species in terms of its number fraction $c_i = N_i/N$, which from Eq. (3.5) can be related to its number density n_i as

$$c_i = \frac{\nu_m}{N_A} n_i \tag{3.6}$$

In terms of concentrations, Eq. (3.5) and Eq. (3.6) give

$$\sum_{i=1}^{n} c_i = 1 \tag{3.7}$$

The Helmoltz free energy is an extensive potential, which implies we can re-write the Helmholtz free energy as $F = V f(T, \{n_i\})$, where f is the free energy density and $\{n_i\}$ is the set of component densities. Similarly, the grand potential can be written as $\Omega = V\omega(T, \{\mu_i\})$, where ω is the grand potential density. The extensive nature of F and Ω thus allow us to write the grand potential density as

$$\omega = f - \sum_{i=1}^{n} \mu_i \, n_i, \tag{3.8}$$

where the chemical potential μ_i is given by

$$\mu_i = \left.\frac{\partial f}{\partial n_i}\right|_{T,V,n_j \neq n_i}, \tag{3.9}$$

From Eq. (3.8), the density counterpart of Eq. (3.4) becomes

$$n_i = -\left.\frac{\partial \omega}{\partial \mu_i}\right|_{T,V,\mu_j \neq \mu_i} \tag{3.10}$$

Equation (3.5) implies that only $n-1$ independent component densities are needed to specify the state of the free energy density (similarly, Eq. (3.7) implies that only $n-1$ concentrations are required to specify the state of the system). Using Eq. (3.5) to eliminate the nth component (we choose this one for simplicity, but we can eliminate any one), we obtain

$$\begin{aligned}\omega &= f_n - \sum_i^{n-1} \mu_i n_i - \mu_n \left(\bar{\rho} - \sum_i^{n-1} n_i\right) \\ &= f_n - \sum_i^{n-1} \left(\mu_i - \mu_n\right) n_i - \mu_n \bar{\rho},\end{aligned} \tag{3.11}$$

where f_n denotes the free energy density with the nth component eliminated in terms of the other $n-1$ densities (i.e. $f_n = f_n(T, \{n_i\}_{n-1})$, where $\{n_i\}_{n-1}$ is the subset of $\{n_i\}$ and f_n is the free energy f with its nth component expressed in terms of the $n_1, n_2, \cdots n_{n-1}$ components.). Switching to concentration variables via Eq. (3.6) and denoting the molar chemical potential (J/mole) by $\tilde{\mu}_i = \bar{\rho}\,\mu_i = (N_A/v_m)\mu_i$ thus gives

$$\omega = f_n - \sum_i^{n-1} \left(\tilde{\mu}_i - \tilde{\mu}_n\right) c_i - \tilde{\mu}_n \tag{3.12}$$

Finally, defining the *inter-diffusion potential* [55] of component i by $\bar{\mu}_i = \tilde{\mu}_i - \tilde{\mu}_n$, and the relative free energy as $\tilde{f}_n = f_n - \tilde{\mu}_n$ gives

$$\omega = \tilde{f}_n - \sum_i^{n-1} \bar{\mu}_i\, c_i \tag{3.13}$$

It is relatively straightforward to show that

$$\left.\frac{\partial \tilde{f}_n}{\partial c_i}\right|_{T,V,c_j \neq c_i} = \bar{\mu}_i \tag{3.14}$$

for $i = 1, 2, \cdots, n-1$, which can be combined with Eq. (3.13) to give

$$\left.\frac{\partial \omega}{\partial \bar{\mu}_i}\right|_{T,V,\bar{\mu}_j \neq \bar{\mu}_i} = -c_i \tag{3.15}$$

We will hereafter simply refer to $\bar{\mu}_i$ as the "chemical potential" of solute c_i (in J/mole), $\tilde{f}_n$ as the free energy density of a phase in terms of the $c_1, c_2, \cdots c_{n-1}$ independent solute concentrations. **For ease of notation in what follows, we drop the tilde n subscript from $\tilde{f}_n$, as well as the over-bar from $\bar{\mu}_i$.**

The next chapters will derive a phase field model of solidification from a grand potential functional expressed as an integral of a grand potential density integrated over the volume V of a solidifying system. The grand potential density will be written in terms of multiple elemental chemical potentials $\{\mu_i\}$ and multiple order parameter fields $\{\phi_\alpha\}$ to represent different orientations and phases in the system.

CHAPTER 4

The Grand Potential Phase Field Functional

THIS CHAPTER CONSTRUCTS THE GRAND potential phase field functional for solidification with multiple order parameters in the spirit of the approach developed by Ofori-Opoku et al. [13]. To proceed, we first clarify some notion and introduce some variables. Let N denote the number of distinct ordered phases or orientations in the system. Define an order parameter vector, $\boldsymbol{\phi}(\vec{r}) = (\phi_1(\vec{r}), \phi_2(\vec{r}), \cdots, \phi_N(\vec{r}))$, the components of which vary from $0 < \phi_i < 1$ $(i = 1, 2, 3, \cdots N)$ and represent the order of one of N solid phases (or orientations) at any location in space, where $\phi_i(\vec{r}) = 0$ represents liquid region and $\phi_i(\vec{r}) > 0$ and ordered regions covered by grain/phase i. Where order parameters interact (e.g. grains merging), they will always be constrained to satisfy $\phi_1 + \phi_2 + \cdots + \phi_N \leq 1$. For an n-component mixture, we define $\boldsymbol{\mu}(\vec{r}) = (\mu_1(\vec{r}), \mu_2(\vec{r}), \cdots, \mu_{n-1}(\vec{r}))$ and $\mathbf{c}(\vec{r}) = (c_1(\vec{r}), c_2(\vec{r}), \cdots, c_{n-1}(\vec{r}))$, which are vector fields representing, respectively, $n-1$ independent chemical potentials and impurity concentration fields at any location in space. In what follows, we also define a set of interpolation functions denoted by $g_\alpha(\boldsymbol{\phi}) \equiv g(\phi_\alpha)$ (where α indexes some component of $\boldsymbol{\phi}$), whose specific form used here is chosen such as to satisfy $g(\phi_\alpha) = 1$ when $\phi_\alpha = 1$ (i.e., the component α of $\boldsymbol{\phi}(\vec{r})$ equals one) and $g(\phi_\alpha) = 0$ when $\phi_\alpha = 0$. Note that in this description where $\phi_\alpha = 1$, all $\phi_{\beta \neq \alpha} \equiv 0$.

DOI: 10.1201/9781003204312-4

In terms of these definitions, the following grand potential of a multi-phase and multi-component solid-liquid system is proposed,

$$\Omega[\boldsymbol{\phi},\boldsymbol{\mu}]=\int_V \underbrace{\left\{\omega_{\text{int}}\left(\boldsymbol{\phi},\boldsymbol{\nabla}\boldsymbol{\phi}\right)+\sum_{\alpha=1}^{N} g_\alpha(\boldsymbol{\phi})\omega^\alpha(\boldsymbol{\mu})+\left[1-\sum_{\alpha=1}^{N} g_\alpha(\boldsymbol{\phi})\right]\omega^\ell(\boldsymbol{\mu})\right\}}_{\omega(\boldsymbol{\phi},\boldsymbol{\nabla}\boldsymbol{\phi},\boldsymbol{\mu})} d^3\boldsymbol{r}, \tag{4.1}$$

where $\omega\left(\boldsymbol{\phi},\boldsymbol{\nabla}\boldsymbol{\phi},\boldsymbol{\mu}\right)$ is the grand potential density, defined here for later reference. In Eq. (4.1), the term $\omega_{\text{int}}\left(\boldsymbol{\phi},\boldsymbol{\nabla}\boldsymbol{\phi}\right)$ tracks the free energy changes associated with solid-liquid interfaces and grain boundaries (i.e. merged order parameters). The index α denotes and runs over solid phases or orientations, while l denotes the liquid. The function $\omega^\vartheta(\boldsymbol{\mu})$ ($\vartheta=\alpha$ or $\vartheta=l$) is the *equilibrium* grand potential density of phase ϑ written in terms of the $n-1$ *non-equilibrium* chemical potentials of the vector field $\boldsymbol{\mu}$. The functions $g_\alpha(\boldsymbol{\phi})$ thus interpolate the local grand potential density between phases via the order parameter components ϕ_α, each of which becomes one (and the others zero) in the bulk of the respective phases they represent. It is noted that $\omega^\vartheta(\boldsymbol{\mu})$ is defined everywhere in space via $\boldsymbol{\mu}(\vec{r})$. Further details of the terms in Eq. (4.1) are discussed in the following sections.

Equation (4.1) assumes that each volume element is like a mini open system that is connected to a thermal heat bath with which it can exchange particles, and which does so subject to the local chemical potential $\boldsymbol{\mu}$. Consistent with density functional theories, thermodynamic equilibrium is attained when $\boldsymbol{\mu}=\boldsymbol{\mu}_{\text{eq}}$, which is constant throughout the system and uniquely defines the equilibrium concentration field throughout the system. We will assume a constant temperature when developing the basic theory of each model in what follows. However, this will be supplemented, as shown in the appendices, with provisions for adjusting the basic models for non-isothermal conditions.

It is also noted that the grand potential density in Eq. (4.1) doesn't contain gradients terms in concentration as is usually required in order to incorporate long-range solute interactions. This is because, as will become apparent later, using the chemical potential as an independent field will require that the concentration be an invertible function of chemical potential. The process becomes generally intractable when the solution of the concentration field comes from a differential equation, rather than an algebraic equation as is the case considered here.

4.1 INTERACTION BETWEEN ORDER PARAMETERS

The first term in Eq. (4.1) is $\omega_{\rm int}$ is given by

$$\omega_{\rm int}\left(\boldsymbol{\phi}, \boldsymbol{\nabla}\boldsymbol{\phi}\right) = \sum_{\alpha=1}^{N} \frac{\sigma_\alpha^2}{2} |\boldsymbol{\nabla}\phi_\alpha|^2 + \sum_{\alpha=1}^{N} H_\alpha f_{\rm DW}(\phi_\alpha) + \sum_{\alpha,\beta\neq\alpha} \omega_{\alpha\beta}\, \Psi(\phi_\alpha, \phi_\beta) \tag{4.2}$$

where nominally σ_α are constants that set the solid-liquid interface energies, H_α defines the nucleation energy between solid α and the liquid, and $f_{\rm DW}$ is some double-well potential (with minima at $\phi_\alpha = 0, 1$). The excess energy of a solid-liquid interface is controlled via σ_α and H_α, and can be done independently of temperature (T), as will be discussed further below. Following Refs. [26, 48], we can also make the σ_α more general functions of all the order parameters, although we shall not consider that case here. The polynomial term $\Psi(\phi_\alpha, \phi_\beta) \propto \phi_\alpha^2\, \phi_\beta^2 + \cdots$ describes an interaction energy when different order parameters overlap (e.g. below the solidus or eutectic temperatures). In what follows, we retain only the second order pair-wise term, although other forms can also be added, such as terms of the form $\phi_\alpha^n\, \phi_\beta^n\, \phi_\gamma^n$, where $n = 2, 4, 6$ (even powers due to symmetry). The excess energy of solid-solid boundaries is controlled by $\omega_{\alpha\beta}$. Specifically, a grain boundary is formed by balancing this interaction term against the driving force of solidification that moves the grains toward each other. In this way, the solid-solid grain boundary can be controlled through the strength of $\omega_{\alpha\beta}$ *and* temperature [13]. It is also noted that for typical values of $\omega_{\alpha\beta}$, overlapping grains always satisfy $\phi_1 + \phi_2 + \cdots \phi_N \leq 1$ everywhere in space. Unlike *multi-phase field* models, where the ϕ_i represent volume fractions, in the above model, the ϕ_i are order parameters and as such can represent the amplitude of density waves in a crystal phase along a certain reciprocal lattice direction. As such, the model above only need constrain the order parameters to satisfy $\phi_1 + \phi_2 + \cdots \phi_N \leq 1$. Modifications to the above model can be added to control solid-solid interface energy independently of temperature. This topic will be further discussed later.

4.2 PROPERTIES OF THE SINGLE-PHASE GRAND POTENTIAL $\omega^\vartheta(\boldsymbol{\mu})$

As is evident from Eq. (4.1), the driving force for solidification is proportional to $\omega_{sl} = \omega^\alpha(\boldsymbol{\mu}) - \omega^\ell(\boldsymbol{\mu})$. The grand potential $\omega^\vartheta(\boldsymbol{\mu})$ of the phase

$\vartheta\,(\alpha, l)$ used in Eq. (4.1) is given by the Legendre transform of the free energy of the phase ϑ,

$$\omega^{\vartheta}\left(\boldsymbol{\mu}\right) = f_{\vartheta}\left(c_1, \cdots, c_{n-1}\right) - \sum_{i=1}^{n-1} \mu_i c_i, \tag{4.3}$$

where f_{ϑ} is defined as the equilibrium Helmhotlz free energy function of phase ϑ and where the $\mu_i = \partial f_{\vartheta}/\partial c_i$ give the chemical potential[1] of species i. Equation (4.3) assumes that f_{ϑ} is a convex function of the concentrations, so that we can express the c_i as a unique function of the chemical potentials μ_i. The multi-component generalization of this is a mapping $\vec{\mu} \mapsto c_i(\vec{\mu})$ that maps the $n-1$ chemical potentials of the system into the same number of solute concentrations. Assuming such an invertible mapping exists, Eq. (4.3) can be written as

$$\begin{aligned}\omega^{\vartheta}\left(\boldsymbol{\mu}\right) = {} & f_{\vartheta}\left(c_1\left(\mu_1, \cdots, \mu_{n-1}\right), \cdots, c_{n-1}\left(\mu_1, \cdots, \mu_{n-1}\right)\right) \\ & - \sum_{i=1}^{n-1} \mu_i c_i\left(\mu_1, \cdots, \mu_{n-1}\right),\end{aligned} \tag{4.4}$$

which then makes the grand potential a function of the chemical potentials, its natural variables. The analogue of Eq. (3.15) can be used to obtain concentration c_i in phase ϑ,

$$c_i^{\vartheta} = -\frac{\partial \omega^{\vartheta}(\boldsymbol{\mu})}{\partial \mu_i} \tag{4.5}$$

The new notation c_i^{ϑ} is introduced to remind the reader that the concentration c_i is calculated from the free energy $f^{\vartheta}(c_i)$. In thermodynamic equilibrium, when $\boldsymbol{\mu}^{\mathrm{eq}}$ is substituted into Eq. (4.5), it yields the equilibrium concentrations c_i corresponding to phase ϑ, hereafter denoted by $c_i^{\vartheta(\mathrm{eq})}$.

4.3 CONCENTRATION IN A MULTI-PHASE SYSTEM IN THE GRAND POTENTIAL ENSEMBLE

In a multi-phase system, equilibrium is described by the grand potential functional of Eq. (4.1), the *concentration field* of a species is now defined

[1]As discussed in detail in Chapter 3 these are actually inter-diffusion potentials but simply referred to hereafter as "chemical potentials" for simplicity.

everywhere by the local chemical potential, which is calculated from the functional generalization of Eq. (3.15) applied to Eq. (4.1), namely,

$$\begin{aligned} c_i(\boldsymbol{\phi}, \boldsymbol{\mu}) &= -\frac{\delta \Omega}{\delta \mu_i} \\ &= -\sum_{\alpha}^{N} g_\alpha(\boldsymbol{\phi}) \frac{\partial \omega^\alpha(\boldsymbol{\mu})}{\partial \mu_i} - \left[1 - \sum_{\alpha}^{N} g_\alpha(\boldsymbol{\phi})\right] \frac{\partial \omega^\ell(\boldsymbol{\mu})}{\partial \mu_i} \\ &= \sum_{\alpha}^{N} g_\alpha(\boldsymbol{\phi}) c_i^\alpha(\boldsymbol{\mu}) + \left[1 - \sum_{\alpha}^{N} g_\alpha(\boldsymbol{\phi})\right] c_i^\ell(\boldsymbol{\mu}), \end{aligned} \tag{4.6}$$

It is recalled that $\omega^\vartheta(\boldsymbol{\mu})$ is the *equilibrium* grand potential function of phase $\vartheta(=\alpha, l)$, evaluated at the field value $\boldsymbol{\mu}$. The concentration field $c_i(\boldsymbol{\phi}, \boldsymbol{\mu})$ becomes $c_i^\vartheta(\boldsymbol{\mu})$ (i.e. Eq. 4.5) within the bulk of each phase ϑ, and interpolated between corresponding values of $c_i^\vartheta(\boldsymbol{\mu})$ across interfaces formed by the components of $\boldsymbol{\phi}$. The explicit variation of Ω from which $c_i(\boldsymbol{\phi}, \boldsymbol{\mu})$ is derived is given by Eq. (5.3). It is noted that when $\boldsymbol{\mu} \rightarrow \boldsymbol{\mu}^{\mathrm{eq}}$, Eq. (4.6) expresses an interpolated equilibrium concentration profile between coexisting solid-liquid phases, or solid-solid phases (depending on what equilibrium $\boldsymbol{\mu}^{\mathrm{eq}}$ defines). The $c_i^\vartheta(\boldsymbol{\mu})$ in Eq. (4.6) shall hereafter be referred to as *auxiliary* fields [24].

The strategy of the grand potential formalism defined by the solidification model above is as follows: the form of $\omega^\vartheta(\boldsymbol{\mu})$ in Eq. (4.4) is first found for all phases ϑ by inverting the system of chemical potentials (if possible) to find the phase concentrations $c_i^\vartheta(\boldsymbol{\mu})$. Once derived, the $\omega^\vartheta(\boldsymbol{\mu})$ are substituted into Eq. (4.1) to describe the local grand potential density of an extended multi-phase, multi-component system in terms of chemical potentials. It is noted that the procedure leading to Eq. (4.4) is only practical if the relations for $\mu_i(c)$ are invertible. For example, if $\mu_i(c)$ is non-convex, the mapping from μ_i to c_i^ϑ is not unique. Alternatively, if $\mu(c_i)$ depended on gradients of the concentration, as in the case in spinodal decomposition, it is also not possible to obtain an algebraic mapping.

CHAPTER 5

Phase Field Dynamics: $\{\phi_\alpha, c_i\}$ versus $\{\phi_\alpha, \mu_i\}$ Evolution

IN TRADITIONAL PHASE FIELD MODELS of solidification, the dynamics of the fields ϕ_α are assumed to follow dissipative dynamics that relax a free energy while coupled to the evolution of conserved solute fields c_i. Here, we will employ a different approach where we couple the evolution of the order parameters ϕ_α to the the intensive variables μ_i that enter Ω and whose evolution is derived from the conservation law of c_i. This will prove to have several practical features.

5.1 TIME EVOLUTION IN THE "TRADITIONAL" FIELDS $\{\phi_\alpha\}$ AND $\{c_i\}$

Order parameters are non-conserved internal degrees of freedom in the system. As a result, the dynamics of each order parameter follows non-conserved gradient flow (*model A*) dynamics,

$$\frac{\partial \phi_\alpha}{\partial t} = -M_{\phi_\alpha} \frac{\delta \Omega}{\delta \phi_\alpha} + \xi_\phi, \tag{5.1}$$

where M_{ϕ_α} defines a time scale for the relaxation of ϕ_α. This will play an important role below in freezing out the kinetics of coalesced interface. The stochastic variable ξ_ϕ accounts for thermal fluctuations and follows the standard fluctuation-dissipation theorem [56].

The evolution of each solute species follows mass conservation dynamics driven by a solute flux that formally couples to the chemical

DOI: 10.1201/9781003204312-5

potential of each species. This is given by conserved (*model B*) dynamics,

$$\begin{aligned}\frac{\partial c_i}{\partial t} &= -\boldsymbol{\nabla}\cdot\boldsymbol{J}_{c_i}+\xi_c \\ &= \boldsymbol{\nabla}\cdot\left(\sum_{j}^{n-1} M_{ij}(\boldsymbol{\phi},\boldsymbol{c})\boldsymbol{\nabla}\mu_j\right)+\nabla\cdot\vec{\zeta}. \end{aligned} \tag{5.2}$$

Where the $M_{ij}(\boldsymbol{\phi},\boldsymbol{c})$ are Osanger-type mobility coefficients for mass transport. Their form depends on the phases and concentrations. Their form for some alloy phases has been worked out elsewhere [57]. The noise $\vec{\zeta}$ is a conserved noise flux governing fluctuations in concentration and also satisfies the fluctuation-dissipation theorem [56].

In what follows, we analyze the above equations of motion in the absence of noise; we will return to the topic of stochastic fluctuations in Chapter 8.

5.2 REFORMULATION OF PHASE FIELD MODEL DYNAMICS IN TERMS OF ϕ_α AND μ_i

In the grand potential ensemble, it is the the order parameters ϕ_i and chemical potentials μ_i that are evolved in time as these are the natural variables of the grand potential that change in space and time. This requires that Eq. (5.2) be re-written with the c_i expressed in terms of the set $\{\mu_i\}$, from which we can obtain the variation of Ω with respect to each respective field ϕ_α ($\alpha = 1, 2, \cdots, N$) and μ_i ($i = 1, 2, \cdots, n-1$) is required. This becomes[1]

$$\begin{aligned}\delta\Omega = \int_V \Bigg(\sum_\alpha^N \Bigg\{ & -\sigma_\alpha^2\boldsymbol{\nabla}^2\phi_\alpha + H_\alpha f'_{\rm DW}(\phi_\alpha) \\ & + \sum_{\beta\neq\alpha} 2\,\omega_{\alpha\,\beta}\phi_\alpha\phi_\beta^2 + g'_\alpha(\boldsymbol{\phi})\left[\omega^\alpha(\boldsymbol{\mu})-\omega^\ell(\boldsymbol{\mu})\right]\Bigg\}\,\delta\phi_\alpha \\ + \sum_i^{n-1}\Bigg\{\sum_\alpha^N g_\alpha(\boldsymbol{\phi})\frac{\partial\omega^\alpha(\boldsymbol{\mu})}{\partial\mu_i} & + \left[1-\sum_\alpha^N g_\alpha(\boldsymbol{\phi})\right]\frac{\partial\omega^\ell(\boldsymbol{\mu})}{\partial\mu_i}\Bigg\}\,\delta\mu_i\Bigg)\,d^3\boldsymbol{r}. \end{aligned} \tag{5.3}$$

where primes in Eq. (5.3) refers to differentiation with respect to the component ϕ_α of $\boldsymbol{\phi}$, and so $g'_\alpha(\boldsymbol{\phi}) \equiv g'(\phi_\alpha)$ throughout. Equation (5.3) will be used as a generating functional for the equations of motion in what follows.

[1]The variation can also be done to take into account varying temperature in a straight forward manner and will be done in Appendix (A).

5.2.1 Order Parameter Evolution

The variational of Eq. (5.3) with respect to the order parameters ϕ_α gives the evolution equation for each order parameter as is given by

$$\frac{1}{M_{\phi_\alpha}}\frac{\partial \phi_\alpha}{\partial t} = \sigma_\alpha^2 \nabla^2 \phi_\alpha - H_\alpha f'_{\mathrm{DW}}(\phi_\alpha) - \sum_{\beta \neq \alpha}^{N} 2\omega_{\alpha\beta}\phi_\alpha\phi_\beta^2 - \left[\omega^\alpha(\boldsymbol{\mu}) - \omega^\ell(\boldsymbol{\mu})\right] g'_\alpha(\boldsymbol{\phi}) \tag{5.4}$$

The square brackets on the right-hand side of Eq. (5.4) define the the thermodynamic driving force. At equilibrium, the chemical potentials μ_i $(i = 1, 2, \cdots n-1)$ comprising the vector $\boldsymbol{\mu}$ become constant throughout the system (and the c_i^ϑ fields defined by Eq. 4.5 attain their equilibrium values in each phase). The equilibrium chemical potentials are denoted, component-wise, by $\mu_i = \mu_i^{\mathrm{eq}}$. It is also noted that at equilibrium (stable or metastable), solid and liquid phases satisfy $\omega^\alpha(\mu_1^{\mathrm{eq}}, \cdots, \mu_{n-1}^{\mathrm{eq}}) = \omega^L(\mu_1^{\mathrm{eq}}, \cdots, \mu_{n-1}^{\mathrm{eq}})$. Thus, at equilibrium, the driving force for the evolution of the ϕ_α goes to zero and the order parameters fields decouple from the chemical potential fields (or concentration fields).

5.2.2 Chemical Potential Evolution

Noting from Eq. (4.6) that c_i are functions of ϕ_α and μ_i, the time derivative of c_i becomes

$$\frac{\partial c_i}{\partial t} = \left(\sum_{\alpha}^{N} \frac{\partial c_i}{\partial \phi_\alpha}\frac{\partial \phi_\alpha}{\partial t} + \sum_{j}^{n-1} \frac{\partial c_i}{\partial \mu_j}\frac{\partial \mu_j}{\partial t} \right) \tag{5.5}$$

Generalizing the susceptibility parameter introduced in Ref. [51], we define a generalized susceptibility matrix by

$$X_{ij} \equiv \chi_i(\boldsymbol{\phi}, \mu_j) \equiv \frac{\partial c_i}{\partial \mu_j} = \sum_{\alpha}^{N} g_\alpha(\boldsymbol{\phi}) \frac{\partial c_i^\alpha(\boldsymbol{\mu})}{\partial \mu_j} + \left[1 - \sum_{\alpha}^{N} g_\alpha(\boldsymbol{\phi})\right] \frac{\partial c_i^\ell(\boldsymbol{\mu})}{\partial \mu_j}, \tag{5.6}$$

From Eq. (4.6), it is also found that

$$\frac{\partial c_i}{\partial \phi_\alpha} = g'_\alpha(\boldsymbol{\phi}) \left[c_i^\alpha(\boldsymbol{\mu}) - c_i^\ell(\boldsymbol{\mu})\right] \tag{5.7}$$

Substituting Eqs. (5.6) and (5.7) into Eq. (5.5), the mass transport equations, Eq. (5.2), becomes (dropping the noise term),

$$\frac{\partial \mu_i}{\partial t} = \sum_j^{n-1} \left(X^{-1}\right)_{ij} \times \left[\boldsymbol{\nabla} \cdot \left(\sum_k^{n-1} M_{jk}(\boldsymbol{\phi}, \mathbf{c}) \boldsymbol{\nabla} \mu_k \right) - \sum_\alpha^N g'_\alpha(\boldsymbol{\phi}) \left[c_j^\alpha(\boldsymbol{\mu}) - c_j^\ell(\boldsymbol{\mu}) \right] \frac{\partial \phi_\alpha}{\partial t} \right] \tag{5.8}$$

where $\left(X^{-1}\right)_{ij}$ denotes the (i, j) component of the inverse of the matrix X_{ij}.

Equations (5.4) and (5.8), along with the $c_i^\alpha = \partial \omega^\alpha(\boldsymbol{\mu}) / \partial \mu_i$ and the definitions for $\omega^\vartheta(\boldsymbol{\mu}) = f_\vartheta\left(\{c_i^\vartheta\}\right) - \sum_{i=1}^{n-1} \mu_i c_i^\vartheta$ comprise a complete set of equations for N order parameters (ϕ_α) and $n - 1$ chemical potential fields (μ_i). For one component $(i = 1)$ and constant mobility, it is relatively straightforward for the reader to show that these equations map onto a phase field model for thermal solidification of a pure material.

CHAPTER 6

Re-Casting Phase Field Equations for Quantitative Simulations

FOR PRACTICAL RATES OF SOLIDIFICATION, it is often desirable for phase field simulations to reproduce the results of the standard *sharp interface models* of solidification [58]. This occurs in the limit when there is a clear separation of scales between the interface width (W) and the solute or thermal diffusion fields around a solidifying front. One way to do this in principle is to make the interface width small (denote this $W \ll d_o$, where d_o is the capillary length of the solid-liquid interface as shown by Caginalp [59, 60]). This is not practical however, as the grid resolution and numerical time scales that result would be computationally intractable for numerical simulations at low to even moderate undercooling. A common strategy for making phase field simulations involving this solidification regime is to smear the interfaces of the ϕ_α fields. Doing so in an uncontrolled way however creates spurious kinetics at the interface, excessive levels of solute trapping in bulk phases and alters the flux conservation across interfaces from its classic form due to lateral diffusion and interface stretching. These effects are physically relevant at rapid rates of solidification where interface kinetics across even a microscopic interface control solidification. For slow to moderate rates of solidification, however, these effects are negligible, and should be eliminated, or reduced, when using diffuse interface models to simulate solidification. The work of Refs. [9, 10, 61] has shown that these effects can be countered numerically—at least in two-phase binary alloys—by

DOI: 10.1201/9781003204312-6

adopting special choices of the interpolation function for solute diffusion and the chemical potential, as well as an addition of a so-called *anti-trapping* flux in the mass transport equations. While the latter adoption precludes the phase field equations from being derived from variational derivatives of the grand-potential, it is suitable if all we care about is emulating the appropriate sharp-interface kinetics across the solid-liquid interface. We will implement analogous modifications to the multi-order parameter phase field models derived in the previous chapters.[1] These are discussed as follows.

6.1 NON-VARIATIONAL MODIFICATIONS TO PHASE FIELD EQUATIONS

The first modification referred to above is made by making the replacement

$$g_\alpha(\boldsymbol{\phi}) \rightarrow h_\alpha(\boldsymbol{\phi}) \tag{6.1}$$

for the interpolation function modulating the concentration in Eq. (4.6). Namely, we modify the concentration field to

$$c_i = \sum_\alpha^N h_\alpha(\boldsymbol{\phi}) c_i^\alpha(\boldsymbol{\mu}) + \left[1 - \sum_\alpha^N h_\alpha(\boldsymbol{\phi})\right] c_i^\ell(\boldsymbol{\mu}), \tag{6.2}$$

from which the modified susceptibility matrix Eq.(5.6) becomes

$$X_{ij} \equiv \chi_i(\boldsymbol{\phi}, \mu_j) = \sum_\alpha^N h_\alpha(\boldsymbol{\phi}) \frac{\partial c_i^\alpha(\boldsymbol{\mu})}{\partial \mu_j} + \left[1 - \sum_\alpha^N h_\alpha(\boldsymbol{\phi})\right] \frac{\partial c_i^\ell(\boldsymbol{\mu})}{\partial \mu_j}, \tag{6.3}$$

where $h_\alpha(\boldsymbol{\phi}) \equiv h(\phi_\alpha)$ is a new function that has the same boundary conditions as $g_\alpha(\boldsymbol{\phi})$ in bulk phases.

The second modification is to introduce a new, so-called, *anti-trapping* current for each component solute. This is given by

$$\boldsymbol{J}_i^{\text{at}} = \sum_\alpha^N a_i(\boldsymbol{\phi}) W_\alpha \left[c_i^\ell(\boldsymbol{\mu}) - c_i^\alpha(\boldsymbol{\mu})\right] \partial_t \phi_\alpha \hat{n}_\alpha \tag{6.4}$$

and is added as a source in the diffusion equation for each component.

[1] While these modifications have only strictly been rigorously derived only for two-phase binary alloys (this derivation will be shown explicitly in Appendix B), we will extend them later for multiple phases and multiple components based on empirical reasoning arguments.

Here, $a_i(\boldsymbol{\phi})$ is a function associated with each solute, to be determined later through matched asymptotic analysis, W_α is the interface width, $\partial_t \phi_\alpha$ is the rate of the respective moving solid-liquid interface and $\hat{n}_\alpha = -\boldsymbol{\nabla}\phi_\alpha / |\boldsymbol{\nabla}\phi_\alpha|$ is the unit normal vector pointing into the liquid. The anti-trapping current modifies the mass conservation equation for each component to

$$\frac{\partial c_i}{\partial t} = -\boldsymbol{\nabla} \cdot \left(\boldsymbol{J}_{c_i} + \boldsymbol{J}_i^{\text{at}}\right) \tag{6.5}$$

Substituting Eq. (6.2) and Eq. (6.4) into Eq. (6.5) thus modifies Eq. (5.8) to

$$\begin{aligned}\frac{\partial \mu_i}{\partial t} &= \sum_j^{n-1} \left(X^{-1}\right)_{ij} \\ &\times \left[\boldsymbol{\nabla} \cdot \left(\sum_k^{n-1} M_{jk} \boldsymbol{\nabla} \mu_k + \sum_\alpha^N a_i(\boldsymbol{\phi}) W_\alpha \left[c_j^\ell(\boldsymbol{\mu}) - c_j^\alpha(\boldsymbol{\mu})\right] \frac{\partial \phi_\alpha}{\partial t} \frac{\boldsymbol{\nabla}\phi_\alpha}{|\boldsymbol{\nabla}\phi_\alpha|}\right)\right. \\ &\left. - \sum_\alpha^N h'_\alpha(\boldsymbol{\phi}) \left[c_j^\alpha(\boldsymbol{\mu}) - c_j^\ell(\boldsymbol{\mu})\right] \frac{\partial \phi_\alpha}{\partial t}\right]\end{aligned} \tag{6.6}$$

Following Ref. [48] we generally express the mobility matrix M_{jk} as

$$M_{jk} = \sum_\alpha^N q_\alpha(\boldsymbol{\phi}) D_{ji}^\alpha X_{ik}^\alpha + \left(1 - \sum_\alpha^N q_\alpha(\boldsymbol{\phi})\right) D_{ji}^L X_{ik}^l, \tag{6.7}$$

where D_{ij}^ϑ is the diffusion matrix of phase ϑ, and the repeated indices denote implied summation over the dummy index $i = 1, 2, \cdots n-1$. The functions $q_\alpha(\boldsymbol{\phi}) (\equiv q(\phi_\alpha))$ is any convenient interpolation functions introduced to vary between 0 in the liquid (all $\phi_\alpha = 0$) and 1 in any grain α (i.e., $\phi_\alpha = 1$). We have also defined two new matrices,

$$\begin{aligned} X_{ij}^\alpha &\equiv \frac{\partial c_i^\alpha}{\partial \mu_j} = \frac{1}{\partial^2 f_\alpha / \partial c_i \partial c_j} \\ X_{ij}^l &\equiv \frac{\partial c_i^L}{\partial \mu_j} = \frac{1}{\partial^2 f_L / \partial c_i \partial c_j}. \end{aligned} \tag{6.8}$$

For multi-phase solidification, Eq. (6.6) describes multi-component diffusion. It is coupled to all the order parameter equations Eq. (5.4), re-written here for convenience for the case of grain α,

$$\tau_\alpha \frac{\partial \phi_\alpha}{\partial t} = W_\alpha^2 \nabla^2 \phi_\alpha - f'_{\rm DW}(\phi_\alpha) - w_{obs}\, \phi_\alpha \sum_{\beta \neq \alpha}^{N} \phi_\beta^2 - \hat{\lambda}_\alpha \left[\omega^\alpha(\boldsymbol{\mu}) - \omega^\ell(\boldsymbol{\mu})\right] g'_\alpha(\boldsymbol{\phi}), \tag{6.9}$$

where $\hat{\lambda}_\alpha \equiv 1/H_\alpha$, $W_\alpha = \sigma_\alpha/\sqrt{H_\alpha}$ and $\tau_\alpha \equiv 1/(M_\phi\, H_\alpha)$. Also, to simplify matters here, we assumed that the interaction term is a constant, i.e. $\omega_{\alpha\beta} = \omega_o$, for any ϕ_α and ϕ_β pair, and define a dimensionless interaction parameter, $\omega_{obs} = 2\,\omega_o/H_\alpha$. As we will allude to in Sections 7.3 and 7.4, we can in principle define $\omega_{\alpha\beta}$ separately for any ϕ_α-ϕ_β combination in such a way as to control the excess energy of an α-β interface.

For future reference when discussing the details of the asymptotic analysis that derives the sharp interface limit of the phase field equations Eq. (6.6) and Eq. (6.9), it is instructive to re-cast Eq. (6.6) in the form of Eq. (5.2). This merges the first and last terms, yielding,

$$\frac{\partial c_i}{\partial t} = \nabla \cdot \left(\sum_j^{n-1} M_{ij} \nabla \mu_j + \sum_\alpha^N W_\alpha\, a_i(\boldsymbol{\phi}) \left[c_i^\ell(\boldsymbol{\mu}) - c_i^\alpha(\boldsymbol{\mu})\right] \frac{\partial \phi_\alpha}{\partial t} \frac{\nabla \phi_\alpha}{|\nabla \phi_\alpha|} \right) \tag{6.10}$$

6.2 CHOICE OF INTERPOLATION FUNCTIONS

Thus far, we have only defined the symmetries of the interpolation functions $h(\phi_\alpha)$, $q(\phi_\alpha)$, $f_{DW}(\phi_\alpha)$ and $g(\phi_\alpha)$ appearing in Eqs. (6.2), (6.3), (6.6), (6.7) and (6.9). There are many choices of interpolation functions one can use in phase field models. Here, it will be convenient to define a working form of these interpolation functions that are commonly used in the literature and which we will use throughout to discuss applications of the phase field models developed. For the case where $0 < \phi_\alpha < 1$, which we have been considering thus far, a convenient choice of interpolation functions is given by

$$\begin{aligned} g(\phi_\alpha) &= \phi_\alpha^3 \left(10 - 15\phi_\alpha + 6\phi_\alpha^2\right) \\ f_{\rm DW}(\phi_\alpha) &= \phi_\alpha^2 \left(1 - \phi_\alpha\right)^2 \\ h(\phi_\alpha) &= \phi_\alpha \\ q(\phi_\alpha) &= \phi_\alpha \\ a_i(\boldsymbol{\phi}) &= = -\frac{1}{\sqrt{2}} \end{aligned} \tag{6.11}$$

We will also be comparing phase field models derived herein with those in the literature, which often use order parameters that vary from $-1 < \phi_\alpha < 1$. A specific set of interpolation functions for these order parameter limits is given by

$$
\begin{aligned}
g(\phi_\alpha) &= \frac{15}{16}\left(\phi_\alpha - \frac{2\phi_\alpha^3}{3} + \frac{\phi_\alpha^5}{5}\right) + \frac{1}{2} \\
f_{\mathrm{DW}}(\phi_\alpha) &= -\frac{\phi_\alpha^2}{2} + \frac{\phi_\alpha^4}{4} \\
h(\phi_\alpha) &= \frac{(\phi_\alpha + 1)}{2} \\
\sum_\alpha^N q_\alpha(\boldsymbol{\phi}) &\rightarrow \frac{(1-\psi)}{2}, \qquad \psi = (N-1) + \sum_{\alpha=1}^N \phi_\alpha \\
a_i(\boldsymbol{\phi}) &= -\frac{1}{2\sqrt{2}}
\end{aligned}
\tag{6.12}
$$

It is noted that the form of the anti-trapping functions $a_i(\boldsymbol{\phi})$ given above are only valid for binary alloys and for the model specializations that we will be working with in later chapters. Their general form cannot be deduced without knowing more about the solidification conditions. In general, their magnitude is chosen such that the anti-trapping current can be used to control the level of solute trapping of each component and other spurious effects caused by the finite (not sharp) thickness of the interface. This topic will be discussed further in later chapters when we also discuss how to choose the interface width and coupling constant $\hat{\lambda}_\alpha$ so as to map special cases of the phase field equations onto the appropriate sharp interface limit.

CHAPTER 7

Equilibrium Properties of the Grand Potential Functional

ONE OF THE MOST IMPORTANT features of *quantitative* phase field models is that the equilibrium order parameter profiles decouple from concentration, or any other diffusing field to which the order parameters are coupled. This makes it possible to specify interface energy independently of the solute profiles through the interface, a practical feature for incorporating the results of microscopic studies of surface energy [17, 62–64]. This decoupling also eliminates the cap on the size of the diffuse interface that can be used, which is numerically expedient [10, 13, 49]. These equilibrium properties are derived in this chapter.

7.1 EQUILIBRIUM CONCENTRATION FIELD

The local *non-equilibrium* concentration of each species in the multi-phase, multi-component system is given by Eq. (4.6). However, in the non-variational form of the phase field equations developed for quantitative applications, this is updated to Eq. (6.2). Two-phase equilibrium between liquid and/or solid phases (whose crystals are indexed by the variable $\alpha = 1, 2, 3, \cdots$) are still defined when the chemical potentials reach a constant, i.e., $\boldsymbol{\mu} = \boldsymbol{\mu}^{\text{eq}}$. The equilibrium concentration field of

DOI: 10.1201/9781003204312-7

species "i" in the system is then given by

$$c_i^{\mathrm{eq}}(\boldsymbol{\phi}) = \sum_{\alpha}^{N} h_\alpha(\boldsymbol{\phi}) c_i^{\alpha(\mathrm{eq})} + \left[1 - \sum_{\alpha}^{N} h_\alpha(\boldsymbol{\phi})\right] c_i^{l(\mathrm{eq})}, \tag{7.1}$$

where

$$c_i^{\vartheta(\mathrm{eq})} = \left.\frac{\partial \omega^\alpha(\boldsymbol{\mu})}{\partial \mu_i}\right|_{\boldsymbol{\mu}=\boldsymbol{\mu}^{\mathrm{eq}}} \equiv c_i^\vartheta(\boldsymbol{\mu} = \boldsymbol{\mu}^{\mathrm{eq}}) \tag{7.2}$$

Here $c_i^{\vartheta(\mathrm{eq})}$ denotes the bulk equilibrium concentration of species "i"in phase ϑ corresponding to the equilibrium chemical potentials $\mu_i = \mu_i^{\mathrm{eq}}$, $\forall i$. It is noted that in multi-component alloys, each μ_i^{eq} in general depends on the nominal (average) alloy composition of each component, $\langle c_i \rangle$, $i = 1, \cdots, n-1$. Equation (7.1) is also applicable to equilibrium between more than two phases. Through Eq. (7.1), the equilibrium concentration field c_i^{eq} is a function of space *only* through the order parameters, a manifestation of the decoupling of order parameters and solute fields at equilibrium in the grand potential formulation.

7.2 EQUILIBRIUM SOLID-LIQUID INTERFACES

To determine the interface energy of a solid-liquid interface (i.e., its excess free energy), consider solid-liquid coexistence along a planar 1D front. This involves a single-phase α and the liquid. In equilibrium, all chemical potentials μ_i are constant everywhere throughout the system. As a result, the grand potential density functions of solid and liquid must be equal, i.e., $\omega^\alpha(\boldsymbol{\mu}^{\mathrm{eq}}) = \omega^\ell(\boldsymbol{\mu}^{\mathrm{eq}})$. Under these conditions, the Euler-Lagrange equation for the steady state equilibrium profile ϕ_α^o becomes

$$\sigma_\alpha^2 \partial_x^2 \phi_\alpha^o - H_\alpha f'_{\mathrm{DW}}(\phi_\alpha^o) = 0 \tag{7.3}$$

The solution of Eq. (7.3) gives the equilibrium solid-liquid interface of the order parameter ϕ_α which satisfies

$$\frac{\partial \phi_\alpha^o}{\partial x} = \frac{\sqrt{H_\alpha}}{\sigma_\alpha} \sqrt{2 f_{\mathrm{DW}}(\phi_\alpha^o)}. \tag{7.4}$$

For a simple ϕ-4 theory, Eq. (7.4) gives a profile described by the convenient hyperbolic tangent function. The interface energy is defined as the excess free energy of the 1D profile according to[1]

$$\gamma_{sl} = \int \{\omega\left(\boldsymbol{\phi}^o, \partial_x \boldsymbol{\phi}^o\right) - \omega\left(\boldsymbol{\phi}^o_{\text{bulk}}\right)\} dx, \tag{7.5}$$

Substituting the square of Eq. (7.4) for the gradient terms in $\omega\left(\boldsymbol{\phi}^o, \partial_x \boldsymbol{\phi}^o\right)$ in Eq. (7.5) gives the solid liquid interface energy as

$$\gamma_{sl} = \sigma_\alpha \sqrt{2\, H_\alpha} \int_0^1 \sqrt{f_{\text{DW}}(\phi^o_\alpha)}\, d\phi_\alpha \tag{7.6}$$

Equation (7.6) reveals that the solid-liquid interface is determined entirely by the choice of σ_α, H_α and f_{DW}, These can also be locally and temporally adjusted in theory to accommodate specific local conditions, although this is rarely done in the literature for simplicity, or simply because it is rarely known for many alloys.

7.3 EQUILIBRIUM SOLID-SOLID INTERFACES: APPROACH I—BASIC MODEL

Two approaches for computing the solid-solid interface energy will be described, the first in this section and the second approach in the following section.

Consider coexistence between two solids, α and β, in a binary two-phase alloy, with a concrete example being a eutectic alloy. Assign each phase described by its own order parameter ϕ_a and ϕ_β. Consider a temperature $T = T_{\text{e}} - \Delta T$, where T_{e} is the eutectic temperature and $\Delta T = T_e - T$. Equilibrium phase coexistence requires that $\omega^\alpha(\boldsymbol{\mu}^{\text{eq}}(T)) = \omega^\beta(\boldsymbol{\mu}^{\text{eq}}(T))$, while at this temperature the liquid is assumed to be *metastable*. Below the eutectic temperature (or the solidus temperature in the case of single-phase solidification), solidification of the liquid into primary or primary and secondary phases can continue until impingement of grains occurs and the solid fraction approaches $f_s \rightarrow 1$. When this occurs, thermodynamic equilibrium is reached throughout the bulks of the system, save for the grain boundaries which are excess free energy structures.

[1]Note that the chemical potential dependency is intentionally left out of $\omega\left(\boldsymbol{\phi}^o, \nabla \boldsymbol{\phi}^o\right)$ in Eq. (7.5) since the driving force for solidification vanishes in solid-liquid equilibrium coexistence.

When impingement occurs the Euler-Lagrange equations describing the steady state profiles of $(\phi^o_\alpha, \phi^o_\beta)$ defining an α-β interface become,

$$\sigma^2_\alpha\, \partial^2_x \phi^o_\alpha - H_\alpha f'_{\text{DW}}(\phi^o_\alpha) - 2\omega_{\alpha\beta}\phi^o_\alpha \left(\phi^o_\beta\right)^2 + \Delta\omega_{sl}\, g'(\phi^o_\alpha) = 0$$
$$\sigma^2_\beta\, \partial^2_x \phi^o_\beta - H_\beta f'_{\text{DW}}(\phi^o_\beta) - 2\omega_{\alpha\beta} \left(\phi^o_\alpha\right)^2 \phi^o_\beta + \Delta\omega_{sl}\, g'(\phi^o_\beta) = 0, \qquad (7.7)$$

where $\Delta\omega_{sl} = \omega^s(\boldsymbol{\mu}^{\text{eq}}(T)) - \omega^L(\boldsymbol{\mu}^{\text{eq}}(T))$ is the solidification driving force, which remains active even below the eutectic (or solidus) temperature in the solids (α or β) and the [meta-stable] liquid. A physically motivated, albeit somewhat loose, interpretation of this is that a solid-solid interface, being of amorphous structure closer to liquid than to solid, persists to be driven toward the crystal state even below the nominal eutectic (or solidus) temperature. Equation (7.7) must be solved numerically to obtain the steady state profiles comprising a solid-solid grain boundary in this model. They are solved subject to the boundary conditions $\phi^o_\alpha(x \to -\infty) = 1$, $\phi^o_\alpha(x \to \infty) = 0$ for the α field, and $\phi^o_\beta(x \to -\infty) = 0$, $\phi^o_\beta(x \to \infty) = 1$ for the β field. It is noteworthy that these equations are explicitly uncoupled from the equilibrium chemical concentration profile $c_i^{\text{eq}}(x)$, making the interface energy of any two-phase combination determinable entirely as a function of temperature (T), $\omega_{\alpha\beta}$, H_α, H_β, σ_α, σ_β. After solving for the steady state profiles, the interface energy across a solid-solid boundary is calculated by substituting these into

$$\gamma_{ss} = \int \{ \omega\left(\boldsymbol{\phi}^o, \partial_x \boldsymbol{\phi}^o, \boldsymbol{\mu}^{\text{eq}}(T)\right) - \omega\left(\boldsymbol{\phi}^o_{\text{bulk}}, \boldsymbol{\mu}^{\text{eq}}(T)\right) \} dx, \qquad (7.8)$$

where here, unlike the case for a solid-liquid interface, the grand potential density difference $\Delta\omega_{sl}$ *does not vanish* at equilibrium below the eutectic or solidus temperatures.

The temperature dependence in the Eq. (7.8) allows control of the temperature dependence of solid-solid interface energies. As shown in Ref [13] this works physically by balancing the driving force for solidification of each order parameter within the overlap regions (since $\Delta\omega_{sl}$ can continue to be non-zero bellow the eutectic or solidus temperatures) against the order parameter interaction that stops different grains from solidifying through one another once a sufficient overlap (ordering) is achieved. This physical interpretation has some physical merit [65], and γ_{ss} can in theory be controlled through the constants in Eq. (7.7). Specifically, the temperature dependence below the solidus can be modulated by changing $\omega_{\alpha\beta}$ (or even σ_α and σ_β). One problem, however is that, except for very low undercooling, the interaction strength $\omega_{\alpha\beta}$ cannot be made arbitrarily small because it leads to unphysical grain overlap and numerically very stiff interfaces due to the large competing driving force on the order parameters. As a result, the above multi-order parameter model, in its present form, only allows for relatively high energy solid-solid boundaries to be modelled.

7.4 EQUILIBRIUM SOLID-SOLID INTERFACES: APPROACH II—MODIFICATION OF MODEL

A more practical approach for modelling grain boundary energies would be to reduce the effect of the temperature-dependent solidification driving force $\Delta\omega_{sl}(\boldsymbol{\mu})$ in Eq. (7.7), making it possible to tune the grain boundary energy to lower values by allowing a lower range of values for the interaction parameter $\omega_{\alpha\beta}$. This can be done by coupling the chemical part of the bulk free energy density in Eq. (4.1) to a new interpolation function that serves to reduce the solidification driving force during grain coalescence. This method was first published in [66] and is reviewed below.

Starting from Eq. (4.1), we multiply all $\omega^{\vartheta}(\boldsymbol{\mu})$ terms by a new interpolation function, denoted $\mathcal{D}(\boldsymbol{\phi}(\boldsymbol{r}))$, that satisfies the following properties: $\mathcal{D}(\boldsymbol{\phi}(\boldsymbol{r}))\rightarrow 0$ wherever $\phi_\alpha\phi_\beta > 0$ for any pair of ϕ_α and ϕ_β (i.e. wherever any grain impingement occurs); $\mathcal{D}(\boldsymbol{\phi}(\boldsymbol{r})) \rightarrow 1$ wherever $\phi_\alpha\phi_\beta = 0$ for any pair of ϕ_α and ϕ_β (i.e. outside of any grain overlap zones); $\delta\mathcal{D}(\boldsymbol{\phi}(\boldsymbol{r}))/\delta\phi_\alpha = 0$ everywhere. Equation (4.1) is thus modified to

$$\Omega[\boldsymbol{\phi},\boldsymbol{\mu}] = \int_V \Bigg\{ \omega_{\text{int}}\left(\boldsymbol{\phi},\boldsymbol{\nabla}\boldsymbol{\phi}\right) + \left[\sum_{\alpha=1}^{N} g_\alpha(\boldsymbol{\phi})\omega^\alpha(\boldsymbol{\mu}) + \left(1-\sum_{\alpha=1}^{N} g_\alpha(\boldsymbol{\phi})\right)\omega^\ell(\boldsymbol{\mu})\right]\mathcal{D}(\boldsymbol{\phi}(\boldsymbol{r}))\Bigg\}\, d^3\boldsymbol{r}. \tag{7.9}$$

Examination of Eq. (7.9) reveals that the grand potential density remains unchanged from its original form everywhere except at solid-solid boundaries formed by overlapping order parameters, where $\omega(\boldsymbol{\phi},\nabla\boldsymbol{\phi},\mu) \rightarrow \omega_{\text{int}}(\boldsymbol{\phi},\nabla\boldsymbol{\phi})$, which is independent of the chemical potential due to the properties of $\mathcal{D}(\boldsymbol{\phi}(\boldsymbol{r}))$, as desired.[2]

[2]This is true in the case of single-phase solidification, where α indexes the same phase, or a multi-phase solidification where the equilibrium chemical potential of multiple solid phases becomes equal in the bulks.

Since $\delta\mathcal{D}(\boldsymbol{\phi}(\boldsymbol{r}))/\delta\phi_\alpha = 0$, the dynamical equations for the order parameters can still be obtained through as variation derivatives of Eq. (7.9), which modifies Eq. (6.9) according to

$$\tau_\alpha \frac{\partial \phi_\alpha}{\partial t} = W_\alpha^2 \nabla^2 \phi_\alpha - f'_{\mathrm{DW}}(\phi_\alpha) - w_{obs}\, \phi_\alpha \sum_{\beta \neq \alpha}^{N} \phi_\beta^2$$
$$- \hat{\lambda}_\alpha \left[\omega^\alpha(\boldsymbol{\mu}) - \omega^\ell(\boldsymbol{\mu})\right] g'_\alpha(\boldsymbol{\phi})\, \mathcal{D}(\boldsymbol{\phi}(\boldsymbol{r})) \qquad (7.10)$$

Moreover, to make the above modification to the model as consistent as possible, the kinetic coefficient τ_α should also be multiplied by an interpolation function analogous to $\mathcal{D}(\boldsymbol{\phi}(\boldsymbol{r}))$ (possibly with a different Λ, so as to simultaneously "freeze out" the kinetics of overlapping order parameters forming a grain boundary. In this modification of the model, *the diffusion of chemical potential will continue to evolve via the non-variational form of Eq. (6.6) already defined in Section 6 and the concentration defined by Eq. (6.2).* This manner of controlling the chemical potential driving force and in tandem with a variational form of the order parameter equations, is designed to provide control over solid-solid grain boundary energies while continuing to maintain the sharp-interface solidification kinetics across solid-liquid interfaces.

With the addition of $\mathcal{D}(\boldsymbol{\phi}(\boldsymbol{r}))$ to grand potential density the equilibrium order parameter profiles defining a stationary flat grain boundary are now given by the simultaneous solution of

$$\sigma_\alpha^2\, \partial_x^2 \phi_\alpha^o - H_\alpha f'_{\mathrm{DW}}(\phi_\alpha^o) - 2\omega_{\alpha\beta}\phi_\alpha^o \left(\phi_\beta^o\right)^2 + \Delta\omega_{sl}\, g'(\phi_\alpha^o)\mathcal{D}_\Lambda(\phi_\alpha, \phi_\beta) = 0$$
$$\sigma_\beta^2\, \partial_x^2 \phi_\beta^o - H_\beta f'_{\mathrm{DW}}(\phi_\beta^o) - 2\omega_{\alpha\beta}\left(\phi_\alpha^o\right)^2 \phi_\beta^o + \Delta\omega_{sl}\, g'(\phi_\beta^o)\mathcal{D}_\Lambda(\phi_\alpha, \phi_\beta) = 0, \qquad (7.11)$$

with the boundary conditions being the same as defined after Eq. (7.7).

We have not yet discussed what form $\mathcal{D}(\boldsymbol{\phi}(\boldsymbol{r}))$ should take. There is a wide class of functional forms we can choose for $\mathcal{D}(\boldsymbol{\phi}(\boldsymbol{r}))$. A simple example is given by

$$\mathcal{D}(\boldsymbol{\phi}(\boldsymbol{r})) = \lim_{\Lambda \to \infty} e^{\left\{-\Lambda \sum_\alpha \sum_\beta \phi_\alpha^2 \phi_\beta^2\right\}}, \qquad (7.12)$$

where the limit is to be taken after evaluating the exponential. It is straightforward to verify that $\mathcal{D}(\boldsymbol{\phi}(\boldsymbol{r}))$ given by Eq. (7.12) satisfies the required properties discussed above[3]. Practically, the function $\mathcal{D}(\boldsymbol{\phi}(\boldsymbol{r}))$ acts similarly to one minus an *indicator function*, which becomes unity anywhere two order parameters do not overlap (i.e. bulk and solid-liquid interfaces) and zero where any order parameter overlap occurs. *Numerically*, it is convenient to define a class of functions denoted $\mathcal{D}_\Lambda(\boldsymbol{\phi}(\boldsymbol{r}))$ given by

$$\mathcal{D}_\Lambda(\boldsymbol{\phi}(\boldsymbol{r})) = e^{\left\{-\Lambda \sum_\alpha \sum_\beta \phi_\alpha^2 \phi_\beta^2\right\}}, \tag{7.13}$$

where Λ is chosen large enough to effectively eliminate the solidification driving force within a reference grain boundary when complete overlap occurs. Specifically, in the limit $\omega_{\alpha\beta} \to 0$, a reference value of Λ can be selected such that $\Delta\omega_{sl}\,\mathcal{D}_\Lambda(\phi_\alpha, \phi_\beta) \to 0$ to some numerical accuracy when overlap occurs, which is defined by $\phi_\alpha + \phi_\beta = 1$ across a grain boundary. It is noted that, while not necessary, for simplicity the interpolation function $g(\phi)$ will be hereafter also be assumed to satisfy the condition $g(\phi_\alpha) + g(\phi_\beta) = 1$ when $\phi_\alpha + \phi_\beta = 1$. This grain overlap criterion and symmetry property of $g(x)$ are straightforward to implement and will be assumed to hold hereafter. This reference boundary corresponds to a neutral grain boundary whose excess energy $\gamma_{ss} = 2\,\gamma_{sl}$, i.e. exactly twice the solid-liquid energy derived in Eq. (7.6) (assuming the constants satisfy $\sigma_\alpha = \sigma_\beta$ and $H_\alpha = H_\beta$).

For non-zero values of $\omega_{\alpha\beta}$, the two order parameters ϕ_α and ϕ_β begin to repel as they approach each other, leading to the condition $\phi_\alpha + \phi_\beta < 1$, which defines the disordered region within a grain boundary, which spans a thickness of approximately $\sim W_\alpha$. As discussed previously, this repulsion sets the scale of the grain boundary energy and balances the solidification driving force that exists within the grain boundary. By multiplying the last term in Eq. (7.10) by $\mathcal{D}_\Lambda(\boldsymbol{\phi}(\boldsymbol{r}))$, however, the thermodynamic driving force for solidification is reduced to a negligible value within the footprint of the grain boundary (i.e. overlapping region of the order parameters). This makes it possible to employ values of $\omega_{\alpha\beta} \sim \mathcal{O}(1)$ or smaller, which in turns allows grain boundary energies that satisfy $2\,\gamma_{sl} < \gamma_{ss} < n\,\gamma_{sl}$, where $n \sim \mathcal{O}(1)$. Of course, given the

[3]Some readers may object that $\phi_\alpha^2 \phi_\beta^2$ is never mathematically zero anywhere. Since this model is destined to be numerical tool, it is always possible to select Λ large enough to make $\mathcal{D}(\boldsymbol{\phi}(\boldsymbol{r}))$ zero within numerical precision only within a suitable overlap region defined by the interface width of the order parameters. It is also noted that there are non-variational forms of $\mathcal{D}(\phi(\mathbf{r}))$ that can be used to control grain boundary energy.

non-linear nature of the correlation of γ_{ss} with $\omega_{\alpha\beta}$, it must be tabulated numerically over a range of temperature by solving Eq. (7.11), which is a relatively simple task.

7.5 SELECTING BETWEEN PHASE FIELD MODELS

The choice of the two model grand potential energy functionals derived above offers different methods for simulating solid-solid interface energies post-solidification. The base model derived form Eq. (4.1) is suitable to use if the user is interested in modelling the growth kinetics and diffusional interactions shaping the morphology of solidifying poly-crystals from a melt. While values of $\omega_{\alpha\beta} \sim \mathcal{O}(10^2)$ with this model will develop generic grain boundaries whose energies are typically quite a bit larger than the normal large-angle grain boundary energies of impinged grains, if the time scale of solidification is relatively rapid compared to the time spent in the completely solid state this should not pose a problem as coarsening kinetics do not change the curvature of the grain boundaries appreciably. If the time spent in the post-solidification state is long enough that solid-state coarsening becomes appreciable, then the second version of the model derived from the grand potential functional in Eq. (7.9) is more suitable to use. However, as discussed above, this version requires some tuning to select a value of Λ (parameter in $\mathcal{D}_\Lambda(\boldsymbol{\phi}(\boldsymbol{r}))$) and to tabulate the equilibrium grain boundary energy as a function of $\omega_{\alpha\beta}$ and temperature.

For the sake of keeping the notation as simple as possible, *the remaining chapter of this book that reference the multi-order parameter phase field model will use the equations defined by the grand potential in Eq. (4.1) and associated dynamical equations given by Eq. (5.4) and Eq. (5.8) as the base model from upon which to chapters build or expand upon. However, it will be implied hereafter that model modifications such as those applied to the phase field equations derived in Section 7.4 (e.g. multiplication of source terms by $\mathcal{D}_\Lambda(\boldsymbol{\phi}(\boldsymbol{r}))$ to decouple solid-solid boundaries from chemical potential) can also be implemented without altering the thermodynamic or kinetic behaviour of any model derived.*

CHAPTER 8

Thermal Fluctuations in Phase Field Equations

TO CAPTURE NUCLEATION AND INTERFACE fluctuations in solidification, phase field equations must introduce stochastic noise into each dynamical equation in order to emulate fast atomic-scale fluctuations that are washed out when one considers dynamical equations driven from a mean-field level free energy. In principle, this can come out formally from direct coarse graining using the methods of statistical mechanics [67], at least for the simplest phase field models. However, when starting directly with phase field models as mesoscale theories in their own right, thermal fluctuations must be added as stochastic noise in the equations of motion, effectively making them Langevin type equations [56]. This section studies how this is done for the phase field equations derived from Eq. 5.4 and Eq. 5.8.

We reproduce here the phase field equations of motion with stochastic noise fields added. For convenience, results are demonstrated for one order parameter and one solute field, and the anti-trapping term is discarded as it will not affect the results discussed in this section. Generalization to multiple fields is straightforward. The order parameter and concentration equations studied here become

$$\tau \frac{\partial \phi}{\partial t} = W_\phi^2 \boldsymbol{\nabla}^2 \phi - f'_{\mathrm{DW}}(\phi) - \hat{\lambda} \left[\omega^\alpha(\mu) - \omega^\ell(\mu) \right] g'(\phi) + \tau \xi, \tag{8.1}$$

$$\frac{\partial c_i}{\partial t} = \boldsymbol{\nabla} \cdot (M \boldsymbol{\nabla} \mu) - \nabla \cdot \vec{\zeta} \tag{8.2}$$

The previously defined constants now become $\hat{\lambda} = 1/H$ and $\tau = 1/(M_\phi H)$, where M_ϕ is the mobility of the order parameter, H is the

DOI: 10.1201/9781003204312-8

nucleation barrier, and M is the solute mobility. The variable ξ is a stochastic scalar field and $\vec{\zeta}$ is a stochastic vector field. Noise sources satisfy the *fluctuation dissipation theorem* [56, 68], namely

$$\langle \xi(\vec{x}, t)\xi(\vec{x}', t')\rangle = 2k_B T M_\phi \,\delta(\vec{x} - \vec{x}')\delta(t - t') \tag{8.3}$$
$$\langle \zeta_i(\vec{x}, t)\zeta_j(\vec{x}', t')\rangle = 2k_B T M \delta(\vec{x} - \vec{x}')\delta(t - t')\delta_{ij}, \tag{8.4}$$

where ζ_i is the ith component of $\vec{\zeta}$ and δ_{ij} is the Kronecker delta.

8.1 NON-DIMENSIONAL FORM OF PHASE FIELD EQUATIONS

It will be useful to work in dimensionless time and space variables. We thus re-couch the above phase field equations and fluctuation dissipation relations in terms of dimensionless time ($\bar{t} = t/\tau$) and space ($\bar{x} = x/W_\phi$). With this rescaling, Eqs. (8.1) and (8.2) become

$$\frac{\partial \phi}{\partial \bar{t}} = \boldsymbol{\nabla}^2 \phi - f'_{\rm DW}(\phi) - \hat{\lambda}\left[\omega^\alpha(\mu) - \omega^\ell(\mu)\right] g'(\phi) + \eta, \tag{8.5}$$
$$\frac{\partial c_i}{\partial \bar{t}} = \boldsymbol{\nabla} \cdot \left(\bar{M} \boldsymbol{\nabla} \mu\right) - \nabla \cdot \vec{q}, \tag{8.6}$$

where the gradients in Eqs. (8.5) and (8.6) are assumed in terms of $\bar{x}$ and where $\eta = \tau\xi$, $\vec{q} = (\tau/W_\phi)\vec{\zeta}$, and $\bar{M} \equiv M\tau/W_\phi^2$. Applying these rescaling to the noise relations in Eqs. (8.3) and (8.4) yields their corresponding dimensionless counterparts,

$$\langle \eta(\vec{\bar{x}}, \bar{t})\eta(\vec{\bar{x}}', \bar{t}')\rangle = 2\frac{k_B T}{H W_\phi^d}\,\bar{\delta}(\vec{\bar{x}} - \vec{\bar{x}}')\bar{\delta}(\bar{t} - \bar{t}') \tag{8.7}$$
$$\langle q_i(\vec{\bar{x}}, \bar{t})q_j(\vec{\bar{x}}', \bar{t}')\rangle = 2\frac{k_B T}{W_\phi^d}\left(\frac{M\tau}{W_\phi^2}\right)\bar{\delta}(\vec{\bar{x}} - \vec{\bar{x}}')\bar{\delta}(\bar{t} - \bar{t}')\delta_{ij}, \tag{8.8}$$

To arrive at Eqs. (8.7) and (8.8), the delta functions were rescaled according to $\delta(\vec{x} - \vec{x}') \to \bar{\delta}(\vec{\bar{x}} - \vec{\bar{x}}')/W_\phi^d$ and $\delta(t - t') \to \bar{\delta}(\bar{t} - \bar{t}')/\tau$, where d is the dimension of space and $\bar{\delta}$ is dimensionless.

It is instructive to re-write the the above noise correlations in terms of the coupling constant between the order parameter and the driving force for solidification. For simplicity, this is done here for the case of a dilute alloy. The form of the noise equations derived will remain valid in the other model specializations examined later.

8.2 SIMPLIFICATION OF NOISE AMPLITUDE FOR THE ORDER PARAMETER EQUATION

To proceed, we first use $RT/\Omega = k_B T \bar{\rho}$, where Ω is the molar volume of the alloy (units of m^3/mole, and assumed to be the same in both phases) and $\bar{\rho}$ is its atomic density (units $\#/m^3$). This is used to re-write Eq. (8.7) as

$$\langle \eta(\vec{x},t)\eta(\vec{x}',t')\rangle = 2\left[\frac{RT(1-k)^2 c_o^L}{\Omega H}\right]\frac{1}{W_\phi^d \bar{\rho}(1-k)^2 c_o^L}\bar{\delta}(\bar{\vec{x}}-\bar{\vec{x}}')\bar{\delta}(\bar{t}-\bar{t}'), \tag{8.9}$$

where k is the partition of the alloy and c_o^L is a reference equilibrium concentration of the liquid. The expression in the square brackets has a special significance, it is related to the dimensional constant

$$\lambda = \frac{15RT(1-k)^2 c_o^L}{16\Omega H}, \tag{8.10}$$

which couples the order parameter and supersaturation in the dilute alloy phase field model of Ref. [10]. It is useful to re-cast the noise correlations in Eq. (8.9) in terms of λ, rather than its specific form for a dilute alloy. Using the above expressions to eliminate W_ϕ in terms of d_o (see Ref. [10] for this relationship) transforms Eq. (8.9) to

$$\langle \eta(\vec{x},t)\eta(\vec{x}',t')\rangle = 2(J\, a_1^d)\left[\frac{F_{\rm exp}}{\lambda^{d-1}}\right]\bar{\delta}(\bar{\vec{x}}-\bar{\vec{x}}')\bar{\delta}(\bar{t}-\bar{t}'), \tag{8.11}$$

where $J = 16/15$, $a_1 = 8839$ (see Ref. [10]) and the dimensionless number $F_{\rm exp}$ is

$$F_{\rm exp} = \frac{\Omega\, k}{N_A\,(1-k)^2\, c_\infty\, d_o^d} \tag{8.12}$$

8.3 SIMPLIFICATION OF NOISE AMPLITUDE FOR THE SOLUTE EQUATION

For the study of fluctuations in the bulk, we expand $\nabla\mu$ as

$$\nabla\mu = \nabla\left(\frac{\partial f}{\partial c}\right) = \frac{\partial \mu}{\partial c}\nabla c, \tag{8.13}$$

from which we identify the solute diffusion coefficient as

$$D = M\frac{\partial \mu}{\partial c} \tag{8.14}$$

For a dilute alloy, $D = (RT/\Omega)\, M/c$. Using $RT/\Omega = k_B T\bar{\rho}$ and noting that $\bar{\rho} = N_A/\Omega$, where N_A is Avogadro's number[1], we can re-write Eq. (8.8)

$$\langle q_i(\bar{\vec{x}}, \bar{t}) q_j(\bar{\vec{x}}', \bar{t}')\rangle = 2\bar{D}\left[\frac{\Omega\, c}{W_\phi^d N_A}\right] \bar{\delta}(\bar{\vec{x}} - \bar{\vec{x}}')\bar{\delta}(\bar{t} - \bar{t}')\, \delta_{ij}, \tag{8.15}$$

where $\bar{D} = D\tau/W_\phi^2$. Equation (8.15) is a dimensionless version of an expression derived for a dilute alloy in Ref. [69]. It is noted that from statistical thermodynamics, the average of $(\delta c)^2$ in a volume ΔV is given by $\langle(\delta c)^2\rangle = \Omega\, c/N_A\, \Delta V$, where $\delta c = c - c_{eq}$, and c_{eq} is the equilibrium concentration[2] of a bulk phase.

At low solidification rates, it will be be shown later that it is possible to re-write the grand potential phase field equations in terms of a *supersaturation field*, which in the bulk becomes $U = (c - c_o^L)/\Delta c_o$, where $\Delta c_o = (1-k)c_o^L = c_\infty(1-k)/k$, with c_∞ being the average alloy concentration and c_o^L is the equilibrium liquid concentrtion at a reference or quench temperature. Transforming Eq. (8.6) dimensionally to the U-field via the substitution $c = \Delta c_o U + c_o^L$ rescales the noise term to $\vec{q}^{\,U} = \vec{q}/\Delta c_o$. Applying this rescaling of the noise to Eq. (8.15), and using $d_o = a_1 W_\phi/\lambda$ (relation between λ and the capillary length d_o, to be discussed later, see Eq. 9.30) to eliminate W_ϕ, gives

$$\langle q_i^U(\vec{x}, t) q_j^U(\vec{x}', t')\rangle = 2\bar{D}\frac{F_{\text{exp}}}{\lambda^d}\left\{1 + (1-k)U\right\} \bar{\delta}(\bar{\vec{x}} - \bar{\vec{x}}')\bar{\delta}(\bar{t} - \bar{t}')\, \delta_{ij}, \tag{8.16}$$

where F_{exp} is given by Eq. (8.12). The average of $(\delta U)^2$ fluctuations in a volume ΔV is thus given by $\langle(\delta U)^2\rangle = d_o F_{\text{exp}}\left\{1 + (1-k)U\right\}/\Delta V$. Equation (8.16) is the dimensionless version of a similar expression derived in Ref. [69].

Eqs. (8.11) and (8.16) are in principle supposed to be applied to each order parameter equation and the chemical potential diffusion equations (or their non-dimensional versions to be precise). It turns out that the

[1]To arrive at this equality, consider N_{tot} atoms that occupy a volume V. If $\bar{\rho}$ is the number density ($\#/m^d$) of the alloy, Ω is its molar volume (m^d/mole) and N_A is Avogadro's number (#/mole), then we can then write $N_{\text{tot}} = V\bar{\rho} = V(N_A/\Omega)$, which gives $\bar{\rho} = (N_A/\Omega)$.

[2]Here, c is expressed in number fraction, n/N where n is the number of solute atoms and N the total number of atoms in a volume V. This is the same as mole fraction, as well as weight fraction *if* the solute and solvent atoms are the same weight.

order parameters noise creates fluctuations of appropriate strength at short wavelengths necessary to promote nucleation, but has a small effect at long wavelengths. On the other hand, the noise associated with the chemical potential creates fluctuations of appropriate strength on long wavelengths and is critical to promote proper side-branching in dendritic growth, but it has a small effect on short wavelengths (see Ref. [69]). For these reasons, it is sometimes appropriate to leave out the chemical potential noise when investigating nucleation or leave out the order parameter noise when investigating solidification morphology.

II

II

CHAPTER 9

Special Cases of the Grand Potential Phase Field Model

THIS CHAPTER SPECIALIZES THE GRAND potential phase field equations Eqs. (6.6) and (6.9) (which were adapted from Eq. 5.4 and Eq. 5.8) to three important case studies of interest. The first deals with the situation of a *multi-component, two-phase* (polycrystalline) alloy where the local grand potential of each phase is assumed to be well approximated by linear deviations in its chemical potentials from its equilibrium value. In other words, this assumes slow enough solidification that the non-equilibrium concentrations don't deviate too much from their equilibrium values. The second case examined in this chapter deals with a *multi-phase binary alloy*, whose local grand potential of each phase can be approximated by a quadratic function away from its equilibrium value. The third case extends this quadratic grand potential approximation to *multi-component* and multi-phase alloys.

9.1 POLYCRYSTALLINE MULTI-COMPONENT ALLOY SOLIDIFICATION AT LOW SUPERSATURATION

Here N order parameters (indexed by α) refer to different orientations of the same solid phase, $\boldsymbol{\mu}$ to the vector of $n-1$ chemical potentials[1] and $\boldsymbol{c}$ the vector of $n-1$ solute concentrations. We proceed by expanding the local grand potential densities the driving force in

[1] Actually, the $n-1$ inter-diffusion potentials as discussed in Chapter 3.

DOI: 10.1201/9781003204312-9

Eq. (6.9) around some equilibrium reference[2] chemical potential denoted by $\boldsymbol{\mu}^{eq} = (\mu_1^{eq}, \mu_2^{eq}, \cdots, \mu_{n-1}^{eq})$. This yields,

$$\begin{aligned} \omega^{\vartheta}(\boldsymbol{\mu}) &= \omega^{\vartheta}(\boldsymbol{\mu}^{eq}) + \sum_i \left.\frac{\partial \omega^{\vartheta}(\boldsymbol{\mu})}{\partial \mu_i}\right|_{\boldsymbol{\mu}^{eq}} (\mu_i - \mu_i^{eq}) \\ &\equiv \omega^{\vartheta}(\boldsymbol{\mu}^{eq}) - \sum_i c_i^{\vartheta(eq)} (\mu_i - \mu_i^{eq}), \end{aligned} \tag{9.1}$$

where Eq. (4.5) was used on the second line. Recalling that $\omega^{\alpha}(\boldsymbol{\mu}^{eq}) = \omega^{L}(\boldsymbol{\mu}^{eq})$ gives

$$\omega^{\alpha}(\boldsymbol{\mu}) - \omega^{L}(\boldsymbol{\mu}) = \sum_i \left(c_i^{l(eq)} - c_i^{\alpha(eq)}\right)(\mu_i - \mu_i^{eq}), \tag{9.2}$$

(where L and l refer to the liquid). We also expand $c_i^L(\boldsymbol{\mu}) - c_i^{\alpha}(\boldsymbol{\mu})$ about $\boldsymbol{\mu}^{eq}$, yielding

$$c_i^L(\mu) - c_i^{\alpha}(\mu) = \Delta c_i + \sum_j \left(X_{ij}^{l(eq)} - X_{ij}^{\alpha(eq)}\right)(\mu_i - \mu_i^{eq}), \tag{9.3}$$

where the concentration differences Δc_i are defined by

$$\Delta c_i = c_i^{l(eq)} - c_i^{s(eq)} \quad \forall\ \alpha \text{ grains}, \tag{9.4}$$

where we have set $c_i^{\alpha(eq)} \equiv c_i^{s(eq)}$, with s denoting the solid phase, while the coefficients $X_{ij}^{\alpha}(\equiv X_{ij}^{s},\ \forall\,\alpha)$ and X_{ij}^{L} are given by

$$\begin{aligned} X_{ij}^{\alpha(\mathrm{eq})} &\equiv \left.\frac{\partial c_i^{\alpha}}{\partial \mu_j}\right|_{\boldsymbol{\mu}^{eq}} \\ X_{ij}^{l(\mathrm{eq})} &\equiv \left.\frac{\partial c_i^{L}}{\partial \mu_j}\right|_{\boldsymbol{\mu}^{eq}} \end{aligned} \tag{9.5}$$

Substituting Eqs. (9.2) and (9.3) into Eqs. (6.6) and (6.9), respectively, yields

$$\begin{aligned} \tau_{\alpha}\frac{\partial \phi_{\alpha}}{\partial t} =& W_{\alpha}^2 \boldsymbol{\nabla}^2 \phi_{\alpha} - f_{\mathrm{DW}}'(\phi_{\alpha}) - w_{obs}\phi_{\alpha}\sum_{\beta\neq\alpha}^{N}\phi_{\beta}^2 \\ & - \hat{\lambda}_{\alpha} g_{\alpha}'(\boldsymbol{\phi}) \sum_j \Delta c_j \left(\mu_j - \mu_j^{\mathrm{eq}}\right) \end{aligned}$$

[2]The manipulations below don't change if we also consider this reference as some non-equilibrium chemical potential.

$$\sum_j X_{ij}\frac{\partial \mu_j}{\partial t} = \nabla\cdot\left\{\sum_{j,k} M_{ij}\,\nabla\mu_j + \left(\Delta c_i + \sum_j\left(X_{ij}^{l(eq)} - X_{ij}^{s(eq)}\right)\left[\mu_j - \mu_j^{\mathrm{eq}}\right]\right)\right.$$
$$\left.\times\sum_\alpha W_\alpha a(\phi)\frac{\partial \phi_\alpha}{\partial t}\frac{\nabla\phi_\alpha}{|\nabla\phi_\alpha|}\right\}$$
$$+\left(\Delta c_i + \sum_j\left(X_{ij}^{l(eq)} - X_{ij}^{s(eq)}\right)\left[\mu_j - \mu_j^{\mathrm{eq}}\right]\right)\sum_\alpha h'(\phi_\alpha)\frac{\partial\phi_\alpha}{\partial t} \tag{9.6}$$

9.1.1 Evaluating the Equilibrium Reference Chemical Potentials $\boldsymbol{\mu}_i^{\mathrm{eq}}$

It is instructive to consider the μ_i^{eq} in an n-component alloy in more detail. Two-phase equilibrium at a given temperature (T) and pressure (p) requires that one of the components of the inter-diffusion potential vector $\boldsymbol{\mu}^{\mathrm{eq}}$ be specified, while the other $n-2$ components of the equilibrium chemical potential difference vector are uniquely determined[3]. This can be understood by the *Gibbs Phase rule* [70], which states that the number of degrees of freedom f (i.e. free thermodynamic variables) over which phase coexistence can be defined is related to the number of components n (i.e. atomic species) and the number of phases r by $f = n - r + 2$. For example, in a ternary alloy $f = 3$; two-phase coexistence at given T and p requires that we also specify one of either $\mu_1^{\mathrm{eq}} = \mu_B^{\mathrm{eq}} - \mu_A^{\mathrm{eq}}$ or $\mu_2^{\mathrm{eq}} = \mu_C^{\mathrm{eq}} - \mu_A^{\mathrm{eq}}$ while the other is uniquely determined. This can be equivalently viewed in terms of concentrations. Two-phase coexistence requires that every liquid concentration pair $(c_B^{l(eq)}, c_C^{l(eq)})$ corresponds to a pair of coexisting solid concentrations $(c_B^{s(eq)}, c_C^{s(eq)})$ through a unique tieline; however, for a given value of $c_C^{l(eq)}$ there is range of $c_B^{l(eq)}$ values for which coexistence can be defined, consistent with the Gibbs phase rule. The degeneracy is broken for a given average concentration pair $(\bar{c}_B, \bar{c}_C)$ in the coexistence region, which connects two unique coexisting pairs $(c_B^{l(eq)}, c_C^{l(eq)})$ with $(c_B^{s(eq)}, c_C^{s(eq)})$, and two corresponding unique values of μ_1^{eq} and μ_2^{eq}.

The considerations of the previous paragraph imply that the equilibrium reference chemical potentials μ_j^{eq} are continuously changing in time and space as the average concentration of a local liquid volume element changes during solidification. Figure (9.1) illustrates this for a portion of

[3]For a binary alloy, $n-2 = 0$, which implies that $\boldsymbol{\mu}^{\mathrm{eq}}$ only contains one component (μ^{eq}), which is uniquely specified at each (T, p) pair.

a dendritic array in a ternary Al-Cu-Si alloy simulated using the model discussed above. The composition maps of $c_{\rm Cu}$ and $c_{\rm Si}$ are shown. In this example $\mu_1^{\rm eq} = \mu_{Cu}^{\rm eq}$ and $\mu_2^{\rm eq} = \mu_{\rm Si}^{\rm eq}$ must be calculated locally in regions such as the ones highlighted by the circles in the figure in order to compute the local driving forces required to numerically advance Eq. (9.6). To determine the $\mu_j^{\rm eq}$ ($j = {\rm Cu, Si}$), Eq. (6.2) is applied to a local volume

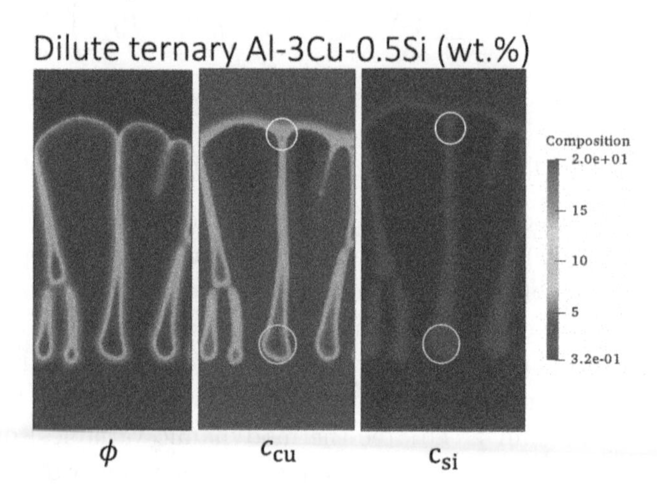

Figure 9.1 A portion of a dendritic array in a ternary Al-Cu-Si alloy, simulated using the model discussed above. The composition maps of $c_{\rm Cu}$ and $c_{\rm Si}$ are shown (as is the order parameter occupying this part of the simulation domain). A pair of circles is shown in the c_{Cu} and c_{Si} frames. These reveal local pockets of microsegregation. (Segregation map is shown in colour in the eBook.)

element having some value of $\boldsymbol{\mu}(\boldsymbol{r})$ to compute its composition and determine if it lies in the single-phase or multi-phase coexistence regions of the alloy phase diagram. In the latter case, the corresponding chemical potentials $\mu_j^{\rm eq}$ must be self-consistently evaluated from a thermodynamic database and input into the local driving forces $\mu_j - \mu_j^{\rm eq}$ in Eq. (9.6). In single-phase regions, the driving terms in the order parameter equations and source terms containing concentrations differences in the chemical potential diffusion equations have no effect because the $g'_\alpha(\phi)$ and $\partial_t \phi_\alpha$ terms vanish. Algorithmically, one can set the local reference chemical potential in the bulks to the local chemical potential itself. It is also noted that the equilibrium concentrations entering Eq. (9.4) can also be evaluated locally once the $\mu_j^{\rm eq}$ is known.

9.1.2 Re-Casting Differential Equations in Eq. (9.6) in Terms of Supersaturation

We next cast the phase field equations in Eq. (9.6) in a simpler and more illuminating form by expressing the driving forces on the right-hand side of both equations in terms of a set of local *supersaturation* fields. These are denoted by U_i and defined by

$$U_i = \frac{X_{ii}^{l(eq)}}{\Delta c_i} \left(\mu_i - \mu_i^{eq}\right), \tag{9.7}$$

for component $i = 1, \cdots, n-1$. We also define a new set of coefficients denoted by k_{ij}^{eff} and defined by

$$k_{ij}^{\text{eff}} \equiv \frac{X_{ij}^{s(eq)}}{X_{ij}^{l(eq)}} \tag{9.8}$$

The coefficients k_{ij}^{eff} form a matrix of ratios of free energy curvatures at the local equilibrium chemical potentials. For the case of an ideal binary alloy, for example, this ratio reduces to the usual partition coefficient. For later use, we re-express the equilibrium limit of the susceptibility derived from Eq. (6.3) in terms of k_{ij}^{eff} as

$$\begin{aligned} X_{ij}^{\text{eq}} \equiv \chi_i(\phi, \mu_j^{eq}) &= \sum_\alpha X_{ij}^{\alpha(eq)}\, h_\alpha(\phi) + X_{ij}^{l(eq)} \left(1 - \sum_\alpha h_\alpha(\phi)\right) \\ &= X_{ij}^{l(eq)} \left\{1 - \left(1 - k_{ij}^{\text{eff}}\right) \sum_\alpha h_\alpha(\phi)\right\} \end{aligned} \tag{9.9}$$

Furthermore, we also define for later use $k_i = c_i^{s(eq)}/c_i^{l(eq)}\ \forall\ \alpha$, which is the equilibrium partition coefficient of solute concentration i. This simplifies the reference equilibrium concentration field in Eq. (7.1) to

$$c_i^{eq}(\phi) = c_i^{l(eq)} \left\{1 - (1 - k_i) \sum_\alpha h_\alpha(\phi)\right\} \tag{9.10}$$

It is recalled, as per the discussion in the last section, that $c_i^{l(eq)}$, k_i, k_{ij}^{eff} and U_i can be computed once the reference chemical potentials μ_i^{eq} are known.

In terms of the variables U_i the first of Eq. (9.6) is re-cast as

$$\tau_\alpha \frac{\partial \phi_\alpha}{\partial t} = W_\alpha^2 \nabla^2 \phi_\alpha - f'_{\text{DW}}(\phi_\alpha) - 2 w_{obs} \phi_\alpha \sum_{\beta \neq \alpha}^{N} \phi_\beta^2 - g'_\alpha(\phi) \sum_{i=1}^{n-1} \lambda_\alpha^i\, U_i, \tag{9.11}$$

where the coupling constants λ_i^α are defined by

$$\lambda_\alpha^i = \frac{\hat{\lambda}_\alpha \Delta c_i^2}{X_{ii}^{l(eq)}}. \tag{9.12}$$

In terms of the U_i and the coefficients k_{ij}^{eff} the second of Eq. (9.6) becomes

$$\begin{aligned}\sum_{j=1}^{n-1} X_{ij}\frac{\partial \mu_j}{\partial t} &= \nabla\cdot\left\{\sum_{j=1}^{n-1} M_{ij}\,\nabla\mu_j + \Delta c_i\left(1+\sum_{j=1}^{n-1}\left(1-k_{ij}^{\text{eff}}\right)\mathcal{R}_{ij}U_j\right)\right.\\ &\quad\left.\times\sum_\alpha W_\alpha\, a_i(\boldsymbol{\phi})\frac{\partial\phi_\alpha}{\partial t}\frac{\nabla\phi_\alpha}{|\nabla\phi_\alpha|}\right\}\\ &\quad + \Delta c_i\left(1+\sum_{j=1}^{n-1}\left(1-k_{ij}^{\text{eff}}\right)\mathcal{R}_{ij}U_j\right)\sum_\alpha h'(\phi_\alpha)\frac{\partial\phi_\alpha}{\partial t},\end{aligned} \tag{9.13}$$

where the matrix $\mathcal{R}_{ij}$ is defined by

$$\mathcal{R}_{ij} \equiv \frac{X_{ij}^{l(eq)}\Delta c_j}{X_{jj}^{l(eq)}\Delta c_i}, \tag{9.14}$$

where *repeated indices in products do not imply summation in Eq. (9.14).* For a binary alloy $\mathcal{R} \to 1$. It is also recalled that mobility tensor M_{ij} is given by Eq. (6.7). We next examine some practical cases of the phase field model defined by Eqs. (9.11) and (9.13).

9.1.3 Practical Limits of Model I: Multi-Component Version of the Model of Ofori-Opoku et al. [13]

Eqs. (9.11) and (9.13) can be reduced to a multi-component form of the model in Ref. [13] in the following limits: the off-diagonal terms in the susceptibility in Eq. (6.8) are neglected and X_{ij}^ϑ are replaced by $X_{ij}^{\vartheta(eq)}$ ($\vartheta = l, \alpha$); the diffusion coefficients in the liquid is assumed constant; the diffusion coefficients in the solid is assumed to be zero (so-called "one-sided" diffusion); this situation is applicable to dilute alloys. These assumptions lead to the following simplifications,

$$
\begin{aligned}
X_{ij}^{\vartheta(eq)} &\to X_{ii}^{\vartheta(eq)}\delta_{ij},\ \vartheta \to \alpha, l \\
D_{ij}^{L} &\to D_{i}^{L}\delta_{ij}, \\
D_{ij}^{\alpha} &= 0 \\
k_{ij}^{\text{eff}} &\approx k_i\delta_{ij} \\
\mathcal{R}_{ij} &\to \delta_{ij},
\end{aligned}
\tag{9.15}
$$

The assumptions in Eq. (9.15) along with Eq. (9.9) reduce Eq. (9.13) to

$$
\begin{aligned}
X_{ii}^{l(eq)}\left\{1-(1-k_i)\sum_{\alpha}h_\alpha(\boldsymbol{\phi})\right\}\frac{\partial \mu_i}{\partial t} &= \boldsymbol{\nabla}\cdot\left(D_i^L\, q_i(\boldsymbol{\phi})\boldsymbol{\nabla}\mu_i\right. \\
&+ \left.\Delta c_i\left\{1+(1-k_i)\,U_i\right\}\sum_{\alpha}W_\alpha a_i(\boldsymbol{\phi})\frac{\partial \phi_\alpha}{\partial t}\frac{\boldsymbol{\nabla}\phi_\alpha}{|\boldsymbol{\nabla}\phi_\alpha|}\right) \\
&+ \Delta c_i\left\{1+(1-k_i)\,U_i\right\}\sum_{\alpha}\frac{\partial h_\alpha(\phi_\alpha)}{\partial t},
\end{aligned}
\tag{9.16}
$$

where $q_i(\boldsymbol{\phi})$ are defined by

$$
\frac{M_{ii}}{D_i^L} \equiv q_i(\boldsymbol{\phi}) = \tilde{q}_i(\boldsymbol{\phi})\,\chi_i(\boldsymbol{\phi}) = X_{ii}^{l(eq)}\,\tilde{q}_i(\boldsymbol{\phi})\left\{1-(1-k_i)\sum_{\alpha}h_\alpha(\boldsymbol{\phi})\right\},
\tag{9.17}
$$

where Eq. (9.17) merely re-casts M_{ii}/D_i^L in Eq. (6.7), and then makes the substitution $q_\alpha(\boldsymbol{\phi}) \to h_\alpha(\boldsymbol{\phi})$. The interpolation function $\tilde{q}_i(\boldsymbol{\phi})$ formally works out to be a function of the sums $\sum_\alpha q_\alpha$ in Eq. (6.7). However, as is done in the literature on quantitative phase field modelling, we are free to treat $\tilde{q}_i(\boldsymbol{\phi})$ as a free degree of freedom and replace it by a convenient interpolation function that varies from 0 in solid and 1 in the liquid. For example, one popular form [10] that we will also use later is

$$
\tilde{q}_i(\boldsymbol{\phi}) = \frac{\bar{q}(\boldsymbol{\phi})}{\{1-(1-k_i)\sum_\alpha h_\alpha(\boldsymbol{\phi})\}}
\tag{9.18}
$$

where $\bar{q}(\boldsymbol{\phi}) = (1-\sum_\alpha \phi_\alpha)$. The order parameter equation Eq. (9.11) remains the same, i.e.

$$
\tau_\alpha\frac{\partial \phi_\alpha}{\partial t} = W_\alpha^2\boldsymbol{\nabla}^2\phi_\alpha - f'_{\text{DW}}(\phi_\alpha) - w_{obs}\phi_\alpha\sum_{\beta\neq\alpha}^{N}\phi_\beta^2 - g'_\alpha(\boldsymbol{\phi})\sum_{i=1}^{n-1}\lambda_\alpha^i\,U_i
\tag{9.19}
$$

where λ_α^i is given by Eq. (9.12).

The derivation of the multi-component version of the model in [13] is made complete by deriving a transformation from U_i to c_i. This is done by expanding c_i near the local equilibrium chemical potentials μ_i^{eq}, yielding

$$c_i = c_i^{eq}(\phi)+\sum_j \left\{ \sum_\alpha \left(h_\alpha(\phi) \left.\frac{\partial c_i^\alpha}{\partial \mu_j}\right|_{\mu_j^{eq}} \right) + \left(1-\sum_\alpha h_\alpha(\phi)\right) \left.\frac{\partial c_i^L}{\partial \mu_j}\right|_{\mu_j^{eq}} \right\} \times \left(\mu_j - \mu_j^{eq}\right), \tag{9.20}$$

where $c_i^{eq}(\phi)$ is given by Eq. (9.10). Equation (9.20) is expressed more compactly as

$$c_i - c_i^{eq}(\phi) = \sum_j \chi_i(\phi, \mu_j^{eq}) \left(\mu_j - \mu_j^{eq}\right) = \sum_j X_{ij}^{eq} \left(\mu_j - \mu_j^{eq}\right) \tag{9.21}$$

where $X_{ij}^{eq} \equiv \chi_i(\phi, \mu_j^{eq})$ (see Eq. 9.9). Combining Eq. (9.21) with Eq. (9.7) yields

$$U_i = \frac{X_{ii}^{l(eq)}}{\Delta c_i} \sum_j \left(X^{eq}\right)_{ij}^{-1} \left\{c_j - c_j^{eq}(\phi)\right\}, \tag{9.22}$$

whose inverse solution gives the solute concentration c_i according to

$$c_i = c_i^{eq}(\phi) + \sum_j \left(\frac{\Delta c_j}{X_{jj}^{l(eq)}} \right) X_{ij}^{eq}\, U_j \tag{9.23}$$

Finally, writing Eq. (9.9) under the simplifications of this section, reduces the supersaturation in Eq. (9.22) to

$$U_i = \frac{c_i - c_i^{eq}(\phi)}{\Delta c_i \left\{1 - (1-k_i)\sum_\alpha h_\alpha(\phi)\right\}} \tag{9.24}$$

For a single component, the special limit considered above reduces exactly to the model of Ref. [13] if the interpolation functions if Eq. (6.12) are used for each order parameter.

9.1.4 Practical Limits of Model II: Two-Phase Binary Alloy Model of Plapp [51]

Eqs. (9.16) and (9.19) also reduce to the two-phase binary alloy model of Ref. [51] for a single component, i.e. if $i = 1$ and we eliminate the sums over α. This leads to: $c_1 \to c$, $U_1 \to U$, $\phi_\alpha \to \phi$, $\Delta c_1 \to \Delta c$, $k_1 \to k^{\text{eff}}$, $X_{11}^{l(eq)} \to \chi^{l(eq)}$, $D_i^L \to D_L$, $\tau_\alpha \to \tau$, $\lambda_\alpha^i \to \lambda$, $h_\alpha \to h$, $g_\alpha \to g$, $q_\alpha \to q$, $N = 1$ and $\sum_\alpha \to 1$. Specifically, this reduces Eqs. (9.16) and (9.19) identically to the supersaturation limit of Ref. [51] [see Eqs. (110)–(112) therein].

9.1.5 Non-Dimensional Form of the Phase Field Model Described by Eqs. (9.16) and (9.19)

This section re-casts Eqs. (9.16) and (9.19) in terms of dimensionless length and time, as well as with the diffusion coefficients scaled to a single reference diffusion coefficient and the driving force in the phase field equations scaled to a single reference coupling constant. This will make them amenable for convenient numerical simulation, and prove convenient to analyze their thin interface limit for quantitative modelling.

To proceed, we assume that $W_\alpha = W_\phi$ and $\tau_\alpha = \tau$ for all order parameters. Next, we scale the diffusion coefficient in each of Eq. (9.16) by the largest diffusion coefficient, which we denote D_f^L (which corresponds to a solute species denoted c_f). Thus, we can write $D_i^L = D_f^L \xi_i$ where $\xi_i = D_i^L / D_f^L$. We also scale the coupling constants in the driving force of Eq. (9.19) by the coupling coefficient λ_α^m corresponding to some *reference* component c_m. Finally, we re-scale time as $\bar{t} = t/\tau$ space with $\bar{x} = x/W_\phi$. With these definitions and scaling, Eq. (9.19) can be re-expressed as

$$\frac{\partial \phi_\alpha}{\partial \bar{t}} = \bar{\boldsymbol{\nabla}}^2 \phi_\alpha - f'_{\text{DW}}(\phi_\alpha) - w_{obs}\phi_\alpha \sum_{\beta \neq \alpha}^{N} \phi_\beta^2 - g'_\alpha(\boldsymbol{\phi})\, \lambda_\alpha^m \left(\sum_{j=1}^{n-1} w_{mj}\, U_j \right), \tag{9.25}$$

where $\bar{\nabla}$ denotes differentiation with respect to $\bar{x}$ (in each dimension). The large bracketed factor in Eq. (9.25) is the total driving force, which is the weighted sum of the U_j, where the weights w_{mj} are given by

$$w_{ij} \equiv \frac{X_{ii}^{l(eq)} \Delta c_j^2}{X_{jj}^{l(eq)} \Delta c_i^2} \tag{9.26}$$

Equation (9.16) is re-cast as

$$X_i^{l(eq)}\left\{1-(1-k_i)\sum_\alpha h_\alpha(\phi)\right\}\frac{\partial\mu_i}{\partial\bar{t}}=\bar{\nabla}\cdot\left(\bar{D}_f\,\xi_i\,q_i(\phi)\bar{\nabla}\mu_i\right.$$
$$+\Delta c_i\left\{1+(1-k_i)\,U_i\right\}\sum_\alpha a_i(\phi)\frac{\partial\phi_\alpha}{\partial\bar{t}}\left.\frac{\bar{\nabla}\phi_\alpha}{|\bar{\nabla}\phi_\alpha|}\right)$$
$$+\Delta c_i\left\{1+(1-k_i)\,U_i\right\}\sum_\alpha\frac{\partial h_\alpha(\phi_\alpha)}{\partial\bar{t}},\tag{9.27}$$

where $\bar{D}_f=D_f^L\tau/W^2$ is a dimensionless reference diffusion coefficient. It is noted that the ratios ξ_i are typically of order 1. The diffusion interpolation function $q_i(\phi)$ is still given by Eq. (9.17), although in the asymptotic analysis discussed below for this variant of the model, it will be convenient to substitute $\hat{q}_i(\phi)=\xi_i q_i(\phi)$ to match the [single-component] form of $q_i(\phi)$ with that used in Appendix (B).

9.1.6 Thin Interface Limit of Phase Field Equations

This section considers how one chooses the parameter relations in the phase field model specializations considered above in order to emulate the kinetics of the classic sharp interface model of solidification that governs slow solidification processes.

Model of Section 9.1.4

For the case of a single-phase, single crystal binary alloy, Eqs. (9.16) and (9.19) reduce exactly to the phase field model in Ref. [51]. In the limit of a diffuse interface, that model was shown to emulate the free-boundary problem of solidification defined by

$$\frac{\partial U}{\partial t}=D_L\nabla^2 U,\quad\text{(liquid)}$$
$$\left[1+(1-k^{\text{eff}})U_{\text{int}}\right]V_n=-D_L\frac{\partial U}{\partial n},$$
$$U_{\text{int}}=-d_o\kappa-\beta V_n,\tag{9.28}$$

where U_{int} is the supersaturation at the sharp interface, d_o is the solutal capillary length and β is the kinetic coefficient. The first equation describes diffusion of solute in the liquid, the second equation expresses flux conservation across the interface and the third equation relates the change of equilibrium concentration as a function of the local interface curvature and velocity. In terms of the concentration field, Eq. (9.28) reduces to the well-known Stefan problem for solidification of a binary alloy,

$$\begin{aligned} \frac{\partial c}{\partial t} &= D_L \nabla^2 c, \quad \text{(liquid)} \\ c|_L\,(1-k^{\text{eff}})v_n &= -D_L \frac{\partial c}{\partial n}\bigg|_L, \\ \delta c|_L &= -\Delta c\left(d_o \kappa + \beta v_n\right), \end{aligned} \tag{9.29}$$

where $\delta c|_L = c|_L - c_L^{eq}$ and $|_L$ here means evaluation on the liquid side of the interface. The coefficients of this Stefan problem are related to phase field model parameters by the relationships

$$\begin{aligned} d_o &= a_1 \frac{W_\phi}{\lambda}, \\ \beta &= a_1 \frac{\tau}{\lambda W_\phi}\left(1 - a_2 \lambda \frac{W_\phi^2}{\tau D_L}\right), \\ \lambda &= \frac{15 \Delta c^2}{16 H \chi^{l(eq)}}, \end{aligned} \tag{9.30}$$

where the constants $a_1 = 0.8839$ and $a_2 = 0.6267$ for the choice of $q(\phi)$ in Eq. (9.17) and the other interpolations functions in Eq. (6.12), and where Δc is the equilibrium concentration jump corresponding to a given quench temperature. *We will hereafter refer to the classic sharp interface model as the special case when $\beta = 0$, which by the second of Eq. (9.30) occurs when $\bar{D} = D_L\tau/W_\phi^2 = a_2\,\lambda$.* The first of Eq. (9.30) is used to set the length scale of the interface W once λ is chosen, while the $\beta = 0$ condition sets the time scale of the phase field kinetics based on the liquid diffusion constant. While λ is theoretically proportional to the inverse of the nucleation barrier, in quantitative modelling it is treated as a numerical convergence parameter, meaning it can be chosen [somewhat] arbitrarily such that simulation results, once scaled back to physical time units using the λ-dependent τ and W do not change. This topic will be discussed further Chapter 11.

Model of Section 9.1.3

We next discuss the thin interface limit of the multi-component, single-crystal limit of Eqs. (9.16) and (9.19). For convenience these will be analyzed in their dimensionless form given by Eqs. (9.25) and (9.27). For multi-component alloys, the sharp interface model becomes

$$
\begin{aligned}
\frac{\partial c_i}{\partial t} &= D_i^L \nabla^2 c_i, \;\; \text{(liquid)} \\
c_i|_L \, (1-k_i) v_n &= -D_i^L \left. \frac{\partial c_i}{\partial n} \right|_L, \\
\sum_{i=1}^{n-1} w_{m\,i} \frac{\delta c_i|_L}{\Delta c_i} &= -d_o \kappa - \beta v_n,
\end{aligned} \tag{9.31}
$$

where $\delta c_i = c_i|_L - c_i^{l(eq)}$ and w_{ij} are given by Eq. (9.26) with the index m denoting a reference solute component. We choose this reference component m to correspond to the scaled coupling constant λ_α^m with which we scaled the driving force in Eq. (9.25). By expanding μ_i to first order in U_i (Eq. 9.7), it is found the left-hand side of the Gibbs-Thomson condition (third line in Eq. 9.31) is just the large bracketed expression in the driving force in Eq. (9.25), evaluated on the liquid side of the interface. These sharp interface conditions can readily be obtained from Eqs. (9.25) and (9.27) in the limit of vanishing interface width W_ϕ by directly analyzing these equations in a curvi-linear co-moving reference frame. In particular, applying the method of projection (see Chapter 4 of [38]) to Eq. (9.25) gives the Gibbs-Thomson condition, while integrating Eq. (6.10) (which gives rise to Eq. 9.27) from $-W_\phi/2 < x < W_\phi/2$ leads to the flux conservation equation (second line in Eq. 9.31).

The so-called "thin interface" or "diffuse interface" analysis of Eqs. (9.25) and (9.27), which is second order accurate in the ratio of W_ϕ/d_o, formally requires a multi-component version of the asymptotic analysis in Appendix (B), where now the spurious interface corrections to the sharp interface model that emerge from the diffuse interface can be eliminated by the choice of anti-trapping flux $a_i(\phi)$, interface diffusion $q_i(\phi)$ as well as the other interpolation functions discussed previously. This analysis is left to the reader as it follows the same steps in Appendix (B). We present a summary in what follows.

We pick up the calculation with Eq. (B.43) for the $\mathcal{O}(\epsilon)$ expansion of the phase field equation. The first term in the expansion of grand potential density difference $\Delta\omega(\mu_0^{\text{in}})$ will now be replaced by a sum over all components. This will will lead to the following modification of Eq. (B.47),

$$
\sum_{i=1}^{n-1} \Delta c_i \, \Delta \mu_{0(i)}^o (0^\pm) = -\bar{D} \alpha \sigma_\phi \bar{v}_0 - \sigma_\phi \alpha \bar{\kappa}, \tag{9.32}
$$

where here Δc_i denotes the equilibrium concentration jump[4] corresponding to component i and $\Delta\mu^o_{0(i)}(0^\pm) = \mu^o_{0(i)}(0^\pm) - \mu_{E(i)}$, where $\mu_{E(i)}$ is the equilibrium chemical potential of component i and $\mu^o_{0(i)}(0^\pm)$ is the lowest order outer chemical potential of component i evaluated on either side of the interface (see Appendix (B) for these and other definitions that follow).

We next turn to the $\mathcal{O}(\epsilon^2)$ expansion for the phase field equation, arriving at Eq. (B.66), the second term of which will now become a sum over components $i = 1, \cdots, n-1$ involving terms of the form $\Delta\tilde{c}_i(\boldsymbol{\mu}_\mathbf{0}^\mathbf{in})\,\mu^{\text{in}}_{1(i)}$, where $\Delta\tilde{c}_i = -\Delta c_i$ and $\boldsymbol{\mu}_\mathbf{0}^\mathbf{in} = (\mu^{\text{in}}_{0(1)}, \cdots, \mu^{\text{in}}_{0(n-1)})$ is the vector of zeroth order corrections to the inner chemical potential of all components, and $\mu^{\text{in}}_{1(i)}$ is the first order correction to the inner chemical potential of component i. We next use the i-component analogues of Eq. (B.51) and Eq. (B.58) to eliminate $\mu^{\text{in}}_{1(i)}$ in terms of the first order correction of the outer chemical potential, $\mu^o_{1(i)}$. This gives the modified version of Eq. (B.69) according to

$$\sum_{i=1}^{n-1} \Delta c_i\, \mu^o_{1(i)}(0^\pm) = -\alpha\bar{D}\sigma_\phi\bar{v}_1 + \left(\sum_{i=1}^{n-1} \Delta c_i\left(K_i + F_i^\pm\right)\right)\bar{v}_0, \tag{9.33}$$

where $0^\pm$ denotes evaluation on either side of the interface. The constants $F_i^\pm$ and K_i are found by adapting the steps in Section B.6.4 of Appendix (B) to the $\mathcal{O}(\epsilon)$ diffusion equation, i.e., start with one equation like Eq. (B.49) for each component. This will lead to the multi-component generalizations of Eqs. (B.59) and (B.68), respectively, which it is noted now formally can include an anti-trapping flux for each component, i.e., $a_t(\phi_0^{\text{in}}) \rightarrow a_{t(i)}(\phi_0^{\text{in}})$. The explicit form of $F_i^\pm$ and K_i will be given below. It is noted that the steps of Section B.6.4 will further lead to the $\mathcal{O}(\epsilon)$ flux conservation condition for each species, however this is not discussed further here[5].

Combining Eq. (9.32) and Eq. (9.33) as was done in the steps leading up to Eq. (B.71) yields a modified $\mathcal{O}(\epsilon^2)$ Gibbs-Thomson condition given

[4]A clarification here. In the notation used in the appendix $\Delta c_i(\boldsymbol{\mu}_\mathbf{0}^\mathbf{in})$ denotes the local interface concentration jump, which is shown to deviate from the corresponding equilibrium value, denoted there as $\Delta c_{F(i)}(\boldsymbol{\mu}_\mathbf{E})$, by a curvature and velocity corrections, which we neglect for reasons discussed in the appendix; in this chapter, the notation Δc_i will continue to denote the equilibrium concentration jump.

[5]Similarly, applying the steps of the analysis of Section B.6.6, the $\mathcal{O}(\epsilon^2)$ diffusion equations (i.e. Eq. B.72) for each component will lead to the next order corrections to the flux conservation conditions, which can be eliminated for each component by the same choice of interpolations functions used here.

by

$$\sum_{i=1}^{n-1} \Delta c_i \Delta\mu_i^o(0^\pm) = -\frac{\sigma_\phi W_\phi}{\hat{\lambda}_\alpha}\kappa - \frac{\tau\sigma_\phi}{W_\phi \hat{\lambda}_\alpha}\left\{1 - \sum_{i=1}^{n-1}\frac{(K_i + F_i^\pm)\,\Delta c_i\,\hat{\lambda}_\alpha}{\sigma_\phi\,\bar{D}}\right\} v_0 \tag{9.34}$$

where $\Delta\mu_i^o(0^\pm) = \mu_i^o(0^\pm) - \mu_{E(i)}$ combines both the first and second order contributions of the asymptotic chemical potential expansion for each component. Equation (9.34) can be clarified further by making the following manipulations. First, we expand the chemical potential to linear order in concentration using $\Delta\mu_i = \Delta c_i / X_{ii}^{l(eq)}$, where $X_{ii}^{l(eq)}$ is given by Eq. (9.5). Next, we substitute the expanded chemical potential into the left-hand side of Eq. (9.34) and multiply both sides of the equation by $X_{ii}^{l(eq)}/\Delta c_m^2$, where m is the chosen index of the reference composition for the analysis of the phase field equations in Section 9.1.5. Finally, we write the coefficients $K_i = \{\Delta c_i/(\xi_i X_{ii}^{l(eq)})\}\bar{K}_i$ and $F_i^\pm = \{\Delta c_i/(\xi_i X_{ii}^{l(eq)})\}\bar{F}_i^\pm$, where $\bar{F}_i^\pm$ and $\bar{K}_i$ become dimensionless. For the choice of $\bar{q}(\phi) = 1 - \phi$ (for all components i) and $h(\phi) = \phi$, the explicit form of $\bar{F}_i^\pm$ and $\bar{K}_i$ is given by Eqs. (13.36) and (13.37), where the only component-dependence comes formally from the modification $a_t(\phi) \to a_t^i(\phi)$. These manipulations are straightforward and are left to the reader[6]. Evaluating the result on the liquid side of the interface gives

$$\sum_{i=1}^{n-1} w_{mi}\frac{\delta c_i(0^+)}{\Delta c_i} = -\frac{\sigma_\phi W_\phi}{\lambda_\alpha^m}\kappa - \frac{\tau\sigma_\phi}{W_\phi \lambda_\alpha^m}\left\{1 - \lambda_\alpha^m \sum_{i=1}^{n-1}\frac{\left(\bar{K}_i + \bar{F}_i\right)}{\sigma_\phi\,\bar{D}\,\xi_i} w_{mi}\right\} v_0, \tag{9.35}$$

where w_{ij} are given by Eq. (9.26) and λ_α^m is given by Eq. (9.12). In writing Eq. (9.35) the substitution $\bar{F}_i^+ = \bar{F}_i^- = \bar{F}_i$ was tacitly assumed, which is required to eliminate solute trapping at the interface, as well as key spurious term arising at second order in the flux conservation condition, i.e. in the i-component analogue of Eq. (B.84). It can be shown that this and the other spurious terms can be eliminated by using $q_i(\phi)$ of the form in Eq. (9.17) and along with the other interpolation functions in Eq. (6.12) for each order parameter. It is also noted that for the case

[6] $\bar{F}^\pm$ and $\bar{K}$ in Eqs. (13.36) and (13.37) are derived from Eqs. (B.59) and (B.68) by scaling out $\Delta c/\bar{\chi}^L$ from their integrands, leading to Eq. (13.32). It is noted that $c_0^{\text{in}}(x)$, $\bar{\chi}(\phi_0^{\text{in}})$ and $\tilde{q}(\phi_0^{in})$ in Eq. (13.33), Eq. (13.34), and Eq. (13.35), respectively, retain the same form in the case of multiple solute components, except that $c_L \to c_L^i$, $k^{\text{eff}} \to k_i^{\text{eff}}$ (taken as $\approx k_i$ here). This leads to $\bar{F}_i^\pm$ and $\bar{K}_i$ that are identical to Eqs. (13.36) and (13.37) with the only i-component dependence formally remaining being that from of the anti-trapping flux.

considered here, the same anti-trapping function (i.e. a constant) can be used for all components, which would make $\bar{K}_i$ and $\bar{F}_i$ in Eq. (9.35) independent of the index i.

Comparing Eq. (9.35) to Eq. (9.31) reveals that the parameter relationships for the effective capillary length and kinetic coefficient corresponding to classic sharp interface limit of the multi-component alloy described by Eqs. (9.25) and (9.27) can be approximated by

$$\begin{aligned} d_o &= a_1 \frac{W_\phi}{\lambda_\alpha^m} \\ \beta &\approx a_1 \frac{\tau}{\lambda_\alpha^m W_\phi}\left(1 - a_2^m \lambda_\alpha^m \frac{W_\phi^2}{\tau D_f^L}\right) \\ \lambda_\alpha^m &= \frac{15\Delta c_m^2}{16 H X_{mm}^{l(eq)}}, \end{aligned} \tag{9.36}$$

where $a_1 = 0.8839$ again, and

$$a_2^m = \sum_{i=1}^{n-1} \left\{ \frac{\left(\bar{K}_i + \bar{F}_i\right)}{\sigma_\phi \, \xi_i} \right\} w_{mi}, \tag{9.37}$$

where the superscript m on a_2^m merely references index in the w_{mi} of Eq. (9.37), which in turn references the component m. As usual, the third expression in Eq. (9.36) is merely formal as λ_α^m can be chosen in Eq. (9.25) as a free convergence parameter that can be made large enough (thus "diffusing" the interface width W_ϕ) such that the final results of the simulations of Eqs. (9.25) and (9.27) do not change within some acceptable accuracy when re-scaled back to dimensional units using τ and W_ϕ.

As in the original binary model explored by Echebarria and coworkers [10], the sum in Eq. (9.37) can be each be modified by a U_i-dependent correction to account for the fact that the kinetic coefficient cannot be made to vanish exactly even up to second order in the asymptotic analysis [10,38]. It is emphasized that the analysis presented here is only valid for a diagonal mobility tensor. A new analysis is required to extract the parametric relationships that would map the full phase field model described by Eqs. (9.11) and (9.13) onto the effective sharp interface model. This is left to future publications.

9.2 MULTI-PHASE BINARY ALLOY WITH QUADRATIC SOLID/LIQUID FREE ENERGIES

This section specializes Eqs. (6.6) and (6.9) to the case of a binary alloy ($i = 1$) with multiple solid phases, each denoted by the index α below. Here we assume that the free energies of any phase can be expanded in the quadratic form

$$f_\vartheta = \frac{A_\vartheta(T)}{2}\left(c - c_\vartheta^{min}(T)\right)^2 + B_\vartheta(T), \quad \vartheta = \alpha, L, \tag{9.38}$$

close to some reference concentration. Here, the solid phases are expanded around their equilibrium concentration with the liquid, while the liquid phase is expanded around its minimum. At the end of this section, we will discuss the situation where the coefficients A_ϑ, c_ϑ^{min} and B_ϑ for the liquid can be defined as piecewise constants in intervals of concentration (or chemical potential) space, thus allowing an accurate description of the free energies over a broad range of concentrations.

9.2.1 Solid-Liquid Phase Coexistence

It is instructive to begin by considering equilibrium between liquid and multiple solid phases, each described by Eq. (9.38). It is assumed that the liquid free energy fit covers a wide enough range of concentrations to simultaneously describe coexistence of the liquid with all solid phases. In terms of the above parabolic free energies, it is straightforward to seek a common tangent that defines the coexistence concentration of the liquid (concentration c_L^{eq}) and a bulk solid phase indexed by α (concentration c_α^{eq}), as well as the corresponding chemical potential μ_α^{eq} between the solid α and the liquid. Such a common tangent is found by solving the system of equations

$$\begin{aligned} \left.\frac{\partial f_L}{\partial c}\right|_{c_L^{eq}} &= \mu_\alpha^{eq} \\ \left.\frac{\partial f_\alpha}{\partial c}\right|_{c_\alpha^{eq}} &= \mu_\alpha^{eq} \\ \frac{f_L - f_\alpha}{c_L^{eq} - c_\alpha^{eq}} &= \mu_\alpha^{eq}, \end{aligned} \tag{9.39}$$

where

$$\mu = \frac{\partial f_\vartheta}{\partial c} = A_\vartheta(c - c_\vartheta^{min}), \quad \vartheta = \alpha, l \tag{9.40}$$

is the chemical potential of each phase at concentration c. The solution of Eq. (9.39) gives

$$c_\vartheta^{eq} = c_\vartheta^{min} + \frac{\mu_\alpha^{eq}}{A_\vartheta}, \quad \vartheta = l, s \tag{9.41}$$

where the equilibrium chemical potential between phase α and liquid is

$$\mu_\alpha^{eq} = \frac{\Delta c_\alpha}{\chi^{l(eq)}\left(1 - k^{\alpha(\text{eff})}\right)}\left(\pm\sqrt{1 + 2\,\frac{\chi^{l(eq)}\left(1 - k^{\alpha(\text{eff})}\right)}{\Delta c_\alpha}\frac{\Delta B_\alpha}{\Delta c_\alpha}} - 1\right), \tag{9.42}$$

where the following definitions were made,

$$\begin{aligned}
\Delta c_\alpha &\equiv c_L^{min} - c_\alpha^{min} \\
\Delta B_\alpha &\equiv B_L - B_\alpha \\
\chi^{l(eq)} &\equiv 1/A_L \\
\chi^{\alpha(eq)} &\equiv 1/A_\alpha \\
k^{\alpha(\text{eff})} &\equiv \chi^{\alpha(eq)}/\chi^{l(eq)}
\end{aligned} \tag{9.43}$$

The sign of the radical in Eq. (9.42) can be chosen such as to match the coexistence of solid and liquid in the appropriate part of concentration space; for example, in a eutectic system, the coexistence of the low-concentration solid with the liquid will have a negative chemical potential above the eutectic temperature, and vice-versa for coexistence of the high-concentration solid with the liquid. (It will be assumed throughout this chapter that the liquid-state coefficients in Eq. 9.43 [e.g. the susceptibilities $\chi^{l(eq)}$] have been pre-determined to fit Eq. 9.38 for coexistence with the corresponding solid phase labelled α.) It is noted that the special limit where $A_L = A_\alpha$ is found by expanding the radical in Eq. (9.42) to leading order in $1 - k^{\alpha(\text{eff})}$, yielding

$$\mu_\alpha^{eq} = \frac{B_L - B_\alpha}{\left(c_L^{min} - c_\alpha^{min}\right)} \tag{9.44}$$

For a practical application of the above free energy fitting procedure applied to Al-Cu system, readers are referred to Ref. [66].

9.2.2 Grand Potential Density of Phase, Multi-Phase Concentration and Susceptibility

The grand potential density of phase ϑ is expressed in terms of μ by eliminating the concentration using Eq. (9.40), namely,

$$\begin{aligned}\omega_\vartheta &= f_\vartheta(c) - \mu c \\ &= -\frac{\mu^2}{2A_\vartheta} - \mu c_\vartheta^{min} + B_\vartheta, \quad \vartheta = \alpha, L,\end{aligned} \tag{9.45}$$

From Eq. (9.45) the concentration of phase ϑ is derived,

$$c_\vartheta = -\frac{\partial \omega_\vartheta}{\partial \mu} = \frac{\mu}{A_\vartheta} + c_\vartheta^{min}, \quad \vartheta = \alpha, L \tag{9.46}$$

In a multi-phase system, Eq. (9.46) can be used to interpolate the local concentration via Eq. (6.2), which gives

$$\begin{aligned}c(\boldsymbol{\phi}, \boldsymbol{\mu}) &= \sum_\alpha h_\alpha(\boldsymbol{\phi}) c_\alpha(\mu) + \left[1 - \sum_\alpha h_\alpha(\boldsymbol{\phi})\right] c_L(\mu) \\ &= \sum_\alpha h_\alpha(\boldsymbol{\phi}) \Big(\frac{\mu}{A_\alpha} + c_\alpha^{min}\Big) + \left[1 - \sum_\alpha h_\alpha(\boldsymbol{\phi})\right] \Big(\frac{\mu}{A_L} + c_L^{min}\Big) \\ &= \chi^{l(eq)} \left\{1 - \sum_\alpha \left(1 - k^{\alpha(\mathrm{eff})}\right) h_\alpha(\boldsymbol{\phi})\right\} \mu - \sum_\alpha \Delta c_\alpha \, h_\alpha(\boldsymbol{\phi}) + c_L^{min},\end{aligned} \tag{9.47}$$

The local equilibrium concentration field in a system comprising a liquid and grains of a solid phase α is thus given by

$$c_\alpha^{\mathrm{eq}}(\boldsymbol{\phi}) = \chi^{l(eq)} \left\{1 - \sum_\alpha \left(1 - k^{\alpha(\mathrm{eff})}\right) h_\alpha(\boldsymbol{\phi})\right\} \mu_\alpha^{\mathrm{eq}} - \sum_\alpha \Delta c_\alpha \, h_\alpha(\boldsymbol{\phi}) + c_L^{min}, \tag{9.48}$$

where here μ_α^{eq} refers to the equilibrium chemical potential between the liquid and the solid phase α with which the liquid is in equilibrium. Finally, from the last line of Eq. (9.47), the generalized susceptibility is derived via the prescription in Eq. (6.3),

$$\chi = \frac{\partial c}{\partial \mu} = \chi^{l(eq)} \left\{1 - \sum_\alpha \left(1 - k^{\alpha(\mathrm{eff})}\right) h_\alpha(\boldsymbol{\phi})\right\} \tag{9.49}$$

It is clear that from Eq. (9.49) that the definition $\chi^{l(eq)} = \chi(\boldsymbol{\phi} = 0)$ and $\chi^{\alpha(eq)}$ is obtained when $\phi_\alpha = 1$ for the α component of $\boldsymbol{\phi}$.

9.2.3 Casting the Chemical Potential and Phase Field Equations in "Supersaturation Form"

By inverting Eq. (9.40) for the liquid phase and the solid phase α, it is found that

$$c^L(\mu) - c^\alpha(\mu) = \Delta c_\alpha + \chi^{l(\mathrm{eq})}\left(1 - k^{\alpha(\mathrm{eff})}\right)\mu \tag{9.50}$$

Meanwhile, from Eq. (9.41) we find

$$\Delta c_\alpha = \Delta c_\alpha^{\mathrm{eq}} - \chi^{l(\mathrm{eq})}\left(1 - k^{\alpha(\mathrm{eff})}\right)\mu_\alpha^{\mathrm{eq}} \tag{9.51}$$

where $\Delta c_\alpha^{\mathrm{eq}} = c_L^{\mathrm{eq}} - c_\alpha^{\mathrm{eq}}$. Substituting Eq. (9.51) into Eq. (9.50), reduces the two-component ($n = 2$) form of Eq. (6.6) for the chemical potential to

$$\begin{aligned}\chi(\boldsymbol{\phi})\frac{\partial \mu}{\partial t} = \boldsymbol{\nabla}\cdot\Bigg[D_L q(\boldsymbol{\phi})\boldsymbol{\nabla}\mu + \sum_\alpha W_\alpha a(\boldsymbol{\phi})\,\Delta c_\alpha^{\mathrm{eq}}\left\{1+\left(1 - k^{\alpha(\mathrm{eff})}\right)U_\alpha\right\}\\ \times\frac{\partial \phi_\alpha}{\partial t}\frac{\boldsymbol{\nabla}\phi_\alpha}{|\boldsymbol{\nabla}\phi_\alpha|}\Bigg]\\ +\frac{1}{2}\sum_\alpha \Delta c_\alpha^{\mathrm{eq}}\left\{1+\left(1 - k^{\alpha(\mathrm{eff})}\right)U_\alpha\right\}\frac{\partial \phi_\alpha}{\partial t},\end{aligned} \tag{9.52}$$

where we have defined the local supersaturation fields U_α for each solid α by

$$U_\alpha \equiv \frac{\chi^{l(\mathrm{eq})}}{\Delta c_\alpha^{\mathrm{eq}}}\left(\mu - \mu_\alpha^{\mathrm{eq}}\right), \tag{9.53}$$

and the binary analogue of the mobility in Eq. (6.7) has been re-shaped into

$$q(\boldsymbol{\phi}) \equiv \frac{M}{D_L} = \sum_\alpha^N q_\alpha(\boldsymbol{\phi})\frac{D^\alpha}{D_L}\chi^{\alpha(eq)} + \left(1 - \sum_\alpha^N q_\alpha(\boldsymbol{\phi})\right)\chi^{l(eq)} \rightarrow \tilde{q}(\boldsymbol{\phi})\,\chi(\boldsymbol{\phi}), \tag{9.54}$$

where the $\rightarrow$ implies that $q(\boldsymbol{\phi})$ has been re-shaped into a free interpolation function $\tilde{q}(\boldsymbol{\phi})$ multiplied by the susceptibility $\chi(\boldsymbol{\phi})$, as in Eq. (9.17), and $\tilde{q}(\boldsymbol{\phi})$ is as the single component version of Eq. (9.18).

To re-cast the order parameter equations, it is noted that using the second line of Eq. (9.45) to write $\omega^\alpha(\mu_\alpha^{\mathrm{eq}}) - \omega^L(\mu_\alpha^{\mathrm{eq}}) = 0$ leads to

$$\Delta B_\alpha = \frac{\chi^{l(\mathrm{eq})}}{2}\left(1 - k^{\alpha(\mathrm{eff})}\right)\left(\mu_\alpha^{\mathrm{eq}}\right)^2 + \Delta c_\alpha \mu_\alpha^{\mathrm{eq}} \tag{9.55}$$

Evaluating $\omega^{\alpha}(\mu) - \omega^{L}(\mu)$ using Eq. (9.45), and using Eq. (9.55), we recast the order parameter equations in Eq. (6.9) for each solid phase α as

$$\tau_\alpha \frac{\partial \phi_\alpha}{\partial t} = W_\alpha^2 \nabla^2 \phi_\alpha - f'_{\mathrm{DW}}(\phi_\alpha) - w_{obs}\phi_\alpha \sum_{\beta\neq\alpha}^{N} \phi_\beta^2$$
$$- \lambda_\alpha U_\alpha \left(1 + \frac{(1 - k^{\alpha(\mathrm{eff})})}{2} U_\alpha\right) g'_\alpha(\phi), \tag{9.56}$$

where

$$\lambda_\alpha \equiv \frac{\hat{\lambda}_\alpha \left(\Delta c_\alpha^{\mathrm{eq}}\right)^2}{\chi^{l(\mathrm{eq})}} \tag{9.57}$$

and where it is recalled that $\hat{\lambda}_\alpha \equiv 1/H_\alpha$ and $\tau_\alpha \equiv 1/(M_{\phi_\alpha} H_\alpha)$.

It is instructive to consider the driving force term of Eq. (9.56) to linear order in U_α. Then, for all orders parameter indices α describing the same solid phase, Eq. (9.52) and Eq. (9.56) identically reduce to the supersaturation limit of Plapp [51], which emulates the sharp interface model described in Eq. (9.28), with capillary length, kinetic coefficient and effective coupling constant given by Eqs. (9.30) (assuming the interpolation functions given by Eq. 6.12). This also is true of the supersaturation limit of Eqs. (6.6) and *(6.9), whose asymptotic analysis is derived in Appendix (B) and yields Eqs. (B.71)* and *(B.84), which form the basis of the Gibbs-Thomson and the flux conservation conditions obeyed asymptotically at the sharp interface defined by each solidifying order parameter.*[7]

When higher order terms in U_α are needed to describe rapid solidification, the full quadratic expression in U_α can be used in the driving force. This term is expected to maintain the linear correspondence between the interface velocity and a non-classical sharp interface model that includes capillarity *and* kinetics, analogously to the approach for pure materials used in Ref. [71]. Strictly speaking, at large solidification speeds, the classical sharp interface model must also be corrected by velocity-dependent partitioning and interface undercooling. These are controlled by solute trapping and drag effects. Chapter 13 will derive such a modified sharp interface model called the *continuous growth model*

[7]The last three terms on the RHS of Eq. (B.84) are corrections must be made to vanish or small in order to emulate the classical sharp-interface model of a binary alloy; this topic is discussed further later.

(CGM), which describes the kinetics of the interface under these conditions. We'll show there how the anti-trapping flux can be used to quantitatively control the level of velocity-dependent solute trapping, without having to resort to the use of intractably small values of the phase field interface width.

9.3 MULTI-PHASE, MULTI-COMPONENT ALLOYS WITH QUADRATIC FREE ENERGIES

This section generalizes the model of Section 9.2 to n components ($n-1$ independent) and N solid phases, indexed with α solidifying from a melt, denoted here by L. Our starting point is again Eqs. (6.6) and (6.9). A phase, in general, will be denoted by ϑ. It is again assumed that the free energy of a phase can be expanded to quadratic order near some reference concentrations. While real materials typically have complex functional forms to describe their free energy, much of the physics of multi-component solidification can be captured with this order of approximation of the free energy [72]. For simplicity, the basic interaction between order parameters will be shown here, however, a more precise control of grain boundaries can easily be achieved by adapting the model's driving force terms and kinetic time scale according to the prescription in Section 7.4.

Note: Readers wishing to skip the algebra (which is a vector version of that done in the last section) can jump directly to the presentation of the phase field equations in Section 9.3.7. **To simplify notation for this section, the remainder of this section will employ a superscript notation to denote the free energy of a phase, i.e. $f_\vartheta \rightarrow f^\vartheta$.**

9.3.1 Free Energy and Susceptibility of a Single Phase

We begin by expressing the free energy of a phase ϑ by the quadratic form

$$f^\vartheta(c_1, c_2, \cdots, c_{n-1}) = \frac{1}{2}\sum_{i=1}^{n-1}\sum_{j=1}^{n-1} A_{ij}^\vartheta \left(c_i - \bar{c}_i^\vartheta\right)\left(c_j - \bar{c}_j^\vartheta\right) + \sum_{j=1}^{n-1} B_j^\vartheta \left(c_j - \bar{c}_j^\vartheta\right) + D^\vartheta, \tag{9.58}$$

where c_i denotes the concentration of component i and A_{ij}^ϑ, $\bar{c}_i^\vartheta$, B_j^ϑ and D_i^ϑ are fitting parameter that are possibly temperature dependent and in principle extractable from thermodynamic databases like Thermo-calc/CALPHAD, at least for select ranges of coexistence between some solid phases and liquid. This is a reasonable form to describe a wide range of solid alloy phases [48, 72]. From Eq. (9.58), the chemical potential in each phase is given by

$$\mu_m^\vartheta = \frac{\partial f^\vartheta}{\partial c_m} = \sum_{j=1}^{n-1} A_{mj}^\vartheta \left(c_j - \bar{c}_j^\vartheta\right) + B_m^\vartheta \tag{9.59}$$

and thus

$$\frac{\partial \mu_m^\vartheta}{\partial c_l} = \frac{\partial^2 f^\vartheta}{\partial c_l \partial c_m} = A_{ml}^\vartheta \tag{9.60}$$

for any concentration c_l. From Eq. (9.60), we define the elements of the inverse susceptibility matrix of a phase by

$$[\chi^\vartheta]_{ij}^{-1} \equiv \frac{\partial \mu_i^\vartheta}{\partial c_j} = A_{ij}^\vartheta. \tag{9.61}$$

For quadratic forms, the matrix $[\boldsymbol{A}^\vartheta]$ is symmetric, which makes its inverse,

$$[\chi^\vartheta] = \begin{bmatrix} \frac{A_{22}^\vartheta}{A_{11}^\vartheta A_{22}^\vartheta - \left(A_{12}^\vartheta\right)^2} & -\frac{A_{12}^\vartheta}{A_{11}^\vartheta A_{22}^\vartheta - \left(A_{12}^\vartheta\right)^2} \\ -\frac{A_{12}^\vartheta}{A_{11}^\vartheta A_{22}^\vartheta - \left(A_{12}^\vartheta\right)^2} & \frac{A_{11}^\vartheta}{A_{11}^\vartheta A_{22}^\vartheta - \left(A_{12}^\vartheta\right)^2} \end{bmatrix} \tag{9.62}$$

also a symmetric matrix.

9.3.2 Vector Notation and Transformations between Concentrations and Chemical Potentials

In what follows, bold letters between square braces will denote $(n-1) \times (n-1)$ matrices and vector arrows will denote $(n-1) \times 1$ column or row matrices. It is noted for future reference that B_m^ϑ shall also be denoted by the alternate form

$$B_m^\vartheta = \mu_m^\vartheta (c_j = \bar{c}_j^\vartheta) \equiv \bar{\mu}_m^\vartheta \tag{9.63}$$

It will prove very useful to couch the multi-order parameter model in matrix notation rather than index notation. An $(n-1) \times 1$ vector is represented by a vertical column of numbers and its transpose by a corresponding horizontal row of numbers. Thus, the $n-1$ concentrations can be equivalently expressed as

$$\vec{c} = \begin{bmatrix} c_1 \\ c_2 \\ \vdots \\ c_{n-1} \end{bmatrix}, \qquad \vec{c}^T = [c_1, c-2, \cdots, c_{n-1}] \tag{9.64}$$

Following this notation, we further define the following vectors to be used below,

$$\vec{\mu}^T = [\mu_1, \mu_2, \cdots, \mu_{n-1}] \tag{9.65}$$

$$(\vec{\bar{c}}^\vartheta)^T = [\bar{c}_1^\vartheta, \bar{c}_2^\vartheta, \cdots, \bar{c}_{n-1}^\vartheta] \tag{9.66}$$

$$(\vec{\bar{\mu}}^\vartheta)^T = [\bar{\mu}_1^\vartheta, \bar{\mu}_2^\vartheta, \cdots, \bar{\mu}_{n-1}^\vartheta] \tag{9.67}$$

$$\Delta\vec{C}^T \equiv (\vec{c}^T - (\vec{\bar{c}}^\vartheta)^T) = [c_1 - \bar{c}_1^\vartheta, c_2 - \bar{c}_2^\vartheta, \cdots, c_{n-1} - \bar{c}_{n-1}^\vartheta] \tag{9.68}$$

$$\Delta\vec{\mu}^T \equiv (\vec{\mu}^T - (\vec{\bar{\mu}}^\vartheta)^T) = [\mu_1 - \bar{\mu}_1^\vartheta, \mu_2 - \bar{\mu}_2^\vartheta, \cdots, \mu_{n-1} - \bar{\mu}_{n-1}^\vartheta] \tag{9.69}$$

In terms of the above definitions, the free energy of phase ϑ (Eq. 9.58) can be written as

$$f^\vartheta(\vec{c}) = \frac{1}{2}\Delta\vec{C}^T[\boldsymbol{\chi}^\vartheta]^{-1}\Delta\vec{C} + \vec{\bar{\mu}}^\vartheta \cdot \Delta\vec{C} + \vec{D}^\vartheta \tag{9.70}$$

where the "$\cdot$" denotes a vector dot product, and where $[\boldsymbol{\chi}^\vartheta]$ is matrix notation for the susceptibility matrix in Eq. (9.61).

Equation (9.59) can be used to define a transformation for commuting between concentration and chemical potential variables, given by

$$\Delta\vec{\mu} = \vec{\mu} - \vec{\bar{\mu}}^\vartheta = [\boldsymbol{\chi}^\vartheta]^{-1}\Delta\vec{C} \tag{9.71}$$

$$\Delta\vec{C} = \vec{c} - \vec{\bar{c}}^\vartheta = [\boldsymbol{\chi}^\vartheta]\,\Delta\vec{\mu}, \tag{9.72}$$

where it is recalled that the components of $\vec{\bar{\mu}}^\vartheta$ are defined by Eq. (9.63).

9.3.3 Grand Potential and Concentration of a Single Phase

The grand potential of phase ϑ is given by

$$\omega^\vartheta = f^\vartheta(c_1, c_2, \cdots, c_{n-1}) - \sum_{j=1}^{n-1} \mu_j\, c_j = f^\vartheta(\vec{c}) - \vec{\mu}\cdot\vec{c} \tag{9.73}$$

Substituting Eq. (9.72) for $\Delta\vec{C}$ in Eq. (9.70), and using Eq. (9.72) to write $\vec{\mu}\cdot\vec{c} = \vec{\mu}^T[\boldsymbol{\chi}^\vartheta]\Delta\vec{\mu} + \vec{\mu}\cdot\vec{\bar{c}}^\vartheta$, tranforms Eq. (9.73) to

$$\omega^\vartheta(\vec{\mu}) = -\frac{1}{2}\Delta\vec{\mu}^T[\boldsymbol{\chi}^\vartheta]\Delta\vec{\mu} - (\vec{\bar{c}}^\vartheta)^T\,\vec{\mu} + \vec{D}^\vartheta \tag{9.74}$$

In arriving at Eq. (9.74), use was made of the interchangeability of the dot product $\vec{\mu}\cdot\vec{\bar{c}}^\vartheta$ with its matrix notation counterpart, i.e.,

$$\vec{\mu}\cdot\vec{\bar{c}}^\vartheta = \vec{\mu}^T\,\vec{\bar{c}}^\vartheta = (\vec{\bar{c}}^\vartheta)^T\,\vec{\mu}, \tag{9.75}$$

which will similarly be assumed for any other dot products that appear below. Equation (9.74) is the vector analougue of Eq. (9.45). From Eq. (9.74) the concentration corresponding to phase ϑ is given by

$$\vec{c}^{\,\vartheta}(\vec{\mu}) = -\frac{\partial\omega^\vartheta}{\partial\vec{\mu}} = [\boldsymbol{\chi}^\vartheta]\Delta\vec{\mu} + \vec{\bar{c}}^\vartheta \tag{9.76}$$

The reader can validate this result by writing Eq. (9.74) in index form and carrying out the differentiation $-\partial\omega^\vartheta/\partial\mu_i$, and similarly for any other equation written in matrix form.

It will be convenient to define the minimum, $\vec{c}_\vartheta^{\,\mathrm{min}}$, of the quadratic form of Eq. (9.58). Setting $\partial f^\vartheta/\partial\vec{c} = 0$ gives

$$\vec{c}_\vartheta^{\,\mathrm{min}} = \vec{\bar{c}}^\vartheta - [\boldsymbol{\chi}^\vartheta]\,\vec{\bar{\mu}}^\vartheta \tag{9.77}$$

Substituting Eq. (9.77) into Eq. (9.76) gives an alternate form for the phase concentration,

$$\vec{c}^{\,\vartheta}(\vec{\mu}) = [\boldsymbol{\chi}^\vartheta]\vec{\mu} + \vec{c}_\vartheta^{\,\mathrm{min}} \tag{9.78}$$

9.3.4 Multi-Phase Concentration, Susceptibility and Concentration Difference

For the case of a multi-phase system, Eq. (6.2) can be used to find the phase-interpolated form of the concentration of each component. This is expressed in vectorial form as

$$\vec{c}(\vec{\phi},\vec{\mu}) = \sum_\alpha h_\alpha(\vec{\phi})\left([\boldsymbol{\chi}^\alpha]\vec{\mu} + \vec{c}_\alpha^{\,\mathrm{min}}\right) + \left\{1 - \sum_\alpha h_\alpha(\vec{\phi})\right\}\left([\boldsymbol{\chi}^L]\vec{\mu} + \vec{c}_L^{\,\mathrm{min}}\right) \tag{9.79}$$

Equation (9.79) can be further simplified by defining

$$\Delta\vec{C}_\alpha = \vec{c}_L^{\,\mathrm{min}} - \vec{c}_\alpha^{\,\mathrm{min}} \tag{9.80}$$

Rearranging Eq. (9.79) and using Eq. (9.80) gives

$$\vec{c}(\vec{\phi},\vec{\mu}) = \left\{[\chi^L] - \sum_\alpha \left([\chi^L] - [\chi^\alpha]\right) h_\alpha(\vec{\phi})\right\}\vec{\mu} - \sum_\alpha \Delta\vec{C}_\alpha\, h_\alpha(\vec{\phi}) + \vec{c}_L^{\,\min} \tag{9.81}$$

Defining an *effective partition matrix* $[\boldsymbol{K}^\alpha]$ by

$$[\boldsymbol{K}^\alpha] = [\chi^L]^{-1}\,[\chi^\alpha], \tag{9.82}$$

the multi-phase concentration can be written as

$$\vec{c}(\vec{\phi},\vec{\mu}) = [\chi^L]\left\{\mathbf{I} - \sum_\alpha \left(\mathbf{I} - [\boldsymbol{K}^\alpha]\right) h_\alpha(\vec{\phi})\right\}\vec{\mu} - \sum_\alpha \Delta\vec{C}_\alpha\, h_\alpha(\vec{\phi}) + \vec{c}_L^{\,\min}, \tag{9.83}$$

where $\mathbf{I}$ denotes the identity matrix. Equation (9.83) is the matrix analogue of Eq. (9.47).

Through the concentration, the susceptibility of a multi-phase system is found by Eq. (6.3), i.e., $\partial c_i/\partial\mu_j$. In vector form, this becomes

$$[\chi] = \frac{\partial\vec{c}}{\partial\vec{\mu}} = [\chi^L]\left\{\mathbf{I} - \sum_\alpha \left(\mathbf{I} - [\boldsymbol{K}^\alpha]\right) h_\alpha(\vec{\phi})\right\} \tag{9.84}$$

Using Eq. (9.78), and making use of Eq. (9.82), we can also compactly express the concentration difference $\vec{c}^{\,L}(\vec{\mu}) - \vec{c}^{\,\alpha}(\vec{\mu})$ between phases as

$$\vec{c}^{\,L}(\vec{\mu}) - \vec{c}^{\,\alpha}(\vec{\mu}) = \Delta\vec{C}_\alpha + [\chi^L]\left\{\mathbf{I} - [\boldsymbol{K}^\alpha]\right\}\vec{\mu}, \tag{9.85}$$

which is the matrix analogue of Eq. (9.50). Equation (9.85) will be relevant to the chemical potential diffusion equation.

It is instructive to re-cast Equation (9.85) into a more convenient form as follows. *Consider that we are in a range of temperature such that for each local chemical potential $\vec{\mu}$ (and local set of liquid compositions) the solid phase α can co-exist, stably or meta-stably, with the liquid phase (L).* As such, a set of equilibrium reference[8] chemical potentials $\{\mu^{\rm eq}_{\alpha(n-1)}\}$ can be deduced from these local conditions, as discussed in

[8] As in the discussion in Section 9.1.1, the manipulations below don't change if we take these reference chemical potentials (and corresponding concentrations) to correspond to some non-equilibrium point.

Section 9.1.1. Evaluating the phase concentration in Eq. (9.78) at the set of chemical potentials $\mu^{\mathrm{eq}}_{\alpha(i)}$, $i = 1, \cdots, n-1$ defines the corresponding equilibrium concentrations of the coexisting liquid (L) and solid (α), i.e.,

$$\vec{c}^{\,\vartheta}_{\mathrm{eq}} = [\boldsymbol{\chi}^{\vartheta}]\vec{\mu}^{\,\mathrm{eq}}_{\alpha} + \vec{c}^{\,\mathrm{min}}_{\vartheta}, \qquad \vartheta = L, \alpha \tag{9.86}$$

Writing Eq. (9.86) for liquid (L) and solid (α), and subtracting the two results gives

$$\Delta\vec{C}_{\alpha} = \Delta\vec{C}^{\alpha}_{\mathrm{eq}} - [\boldsymbol{\chi}^{L}]\left\{\mathbf{I} - [\boldsymbol{K}^{\alpha}]\right\}\vec{\mu}^{\,\mathrm{eq}}_{\alpha}, \tag{9.87}$$

where

$$\Delta\vec{C}^{\alpha}_{\mathrm{eq}} \equiv \vec{c}^{\,L}_{\mathrm{eq}} - \vec{c}^{\,\alpha}_{\mathrm{eq}} \tag{9.88}$$

Substituting Eq. (9.87) into Eq. (9.85) and re-arranging reduces $\vec{c}^{\,L}(\vec{\mu}) - \vec{c}^{\,\alpha}(\vec{\mu})$ to

$$\vec{c}^{\,L}(\vec{\mu}) - \vec{c}^{\,\alpha}(\vec{\mu}) = \Delta\vec{C}^{\alpha}_{\mathrm{eq}} + [\boldsymbol{\chi}^{L}]\left\{\mathbf{I} - [\boldsymbol{K}^{\alpha}]\right\}(\vec{\mu} - \vec{\mu}^{\,\mathrm{eq}}_{\alpha}) \tag{9.89}$$

9.3.5 Grand Potential Driving Force for Multi-Phase Solidification

The final quantity we need to derive the phase field equations we seek in this section is the driving force for each order parameter (phase) i.e., $\omega^{L}(\vec{\mu}) - \omega^{\alpha}(\vec{\mu})$. Using Eq. (9.74) to evaluate the grand potential of each phase gives,

$$\begin{aligned}
&\omega^{L}(\vec{\mu}) - \omega^{\alpha}(\vec{\mu}) = \\
&\quad -\frac{1}{2}\vec{\mu}^{\mathrm{T}}[\boldsymbol{\chi}^{L}]\vec{\mu} + \frac{1}{2}\vec{\mu}^{\mathrm{T}}[\boldsymbol{\chi}^{L}]\vec{\bar{\mu}}^{L} + \frac{1}{2}(\vec{\bar{\mu}}^{L})^{\mathrm{T}}[\boldsymbol{\chi}^{L}]\vec{\mu} - \frac{1}{2}(\vec{\bar{\mu}}^{L})^{\mathrm{T}}[\boldsymbol{\chi}^{L}]\vec{\bar{\mu}}^{L} \\
&\quad + \frac{1}{2}\vec{\mu}^{\mathrm{T}}[\boldsymbol{\chi}^{\alpha}]\vec{\mu} - \frac{1}{2}\vec{\mu}^{\mathrm{T}}[\boldsymbol{\chi}^{\alpha}]\vec{\bar{\mu}}^{\alpha} - \frac{1}{2}(\vec{\bar{\mu}}^{\alpha})^{\mathrm{T}}[\boldsymbol{\chi}^{\alpha}]\vec{\mu} + \frac{1}{2}(\vec{\bar{\mu}}^{\alpha})^{\mathrm{T}}[\boldsymbol{\chi}^{\alpha}]\vec{\bar{\mu}}^{\alpha} \\
&\quad - \left(\vec{\bar{c}}^{L} - \vec{\bar{c}}^{\alpha}\right)^{T}\vec{\mu} + \left(\vec{\bar{D}}^{L} - \vec{\bar{D}}^{\alpha}\right)
\end{aligned} \tag{9.90}$$

Several tedious "clean-up" steps are required to simplify Eq. (9.90). The first is to use Eq. (9.77) to write

$$\left(\vec{\bar{c}}^{L} - \vec{\bar{c}}^{\alpha}\right)^{\mathrm{T}}\vec{\mu} = (\Delta\vec{C}_{\alpha})^{T}\vec{\mu} + (\vec{\bar{\mu}}^{L})^{\mathrm{T}}[\boldsymbol{\chi}^{L}]\vec{\mu} - (\vec{\bar{\mu}}^{\alpha})^{\mathrm{T}}[\boldsymbol{\chi}^{\alpha}]\vec{\mu} \tag{9.91}$$

The second is to recall that, by definition, $\omega^L(\vec{\mu}_\alpha^{\rm eq}) = \omega^\alpha(\vec{\mu}_\alpha^{\rm eq})$, which, using Eq. (9.74) allows us to calculate $\vec{D}^L - \vec{D}^\alpha$, yielding

$$\begin{aligned}
&\vec{D}^L - \vec{D}^\alpha = \\
&-\frac{1}{2}(\vec{\mu}_\alpha^{\rm eq})^T[\boldsymbol{\chi}^\alpha]\vec{\mu}_\alpha^{\rm eq} + \frac{1}{2}(\vec{\mu}_\alpha^{\rm eq})^T[\boldsymbol{\chi}^\alpha]\vec{\bar{\mu}}^\alpha + \frac{1}{2}(\vec{\bar{\mu}}^\alpha)^{\rm T}[\boldsymbol{\chi}^\alpha]\vec{\mu}_\alpha^{\rm eq} - \frac{1}{2}(\vec{\bar{\mu}}^\alpha)^{\rm T}[\boldsymbol{\chi}^\alpha]\vec{\bar{\mu}}^\alpha \\
&+\frac{1}{2}(\vec{\mu}_\alpha^{\rm eq})^T[\boldsymbol{\chi}^L]\vec{\mu}_\alpha^{\rm eq} - \frac{1}{2}(\vec{\mu}_\alpha^{\rm eq})^T[\boldsymbol{\chi}^L]\vec{\bar{\mu}}^L - \frac{1}{2}(\vec{\bar{\mu}}^L)^{\rm T}[\boldsymbol{\chi}^L]\vec{\mu}_\alpha^{\rm eq} + \frac{1}{2}(\vec{\bar{\mu}}^L)^{\rm T}[\boldsymbol{\chi}^L]\vec{\bar{\mu}}^L \\
&+\left(\vec{\bar{c}}^L - \vec{\bar{c}}^\alpha\right)^T \vec{\mu}_\alpha^{\rm eq}
\end{aligned} \tag{9.92}$$

When Eq. (9.91) and Eq. (9.92) are inserted in the last line of Eq. (9.90), the expression $\left(\vec{\bar{c}}^L - \vec{\bar{c}}^\alpha\right)^T \vec{\mu}_\alpha^{\rm eq} - (\Delta\vec{C}_\alpha)^T\vec{\mu}$ emerges, which we need to simplify further. The second terms in this expression is derived directly from Eq. (9.86). The first term is found by equating the expression for $\vec{c}_\vartheta^{\,\rm min}$ from Eq. (9.77) and Eq. (9.86). Proceeding thus, this expression becomes,

$$\begin{aligned}
\left(\vec{\bar{c}}^L - \vec{\bar{c}}^\alpha\right)^T \vec{\mu}_\alpha^{\rm eq} - (\Delta\vec{C}_\alpha)^T\vec{\mu} &= -(\vec{\mu}_\alpha^{\rm eq})^T\left([\boldsymbol{\chi}^L] - [\boldsymbol{\chi}^\alpha]\right)\vec{\mu}_\alpha^{\rm eq} \\
&+ (\vec{\mu}_\alpha^{\rm eq})^T\left([\boldsymbol{\chi}^L] - [\boldsymbol{\chi}^\alpha]\right)\vec{\mu} \\
&+ (\vec{\bar{\mu}}^L)^T[\boldsymbol{\chi}^L]\vec{\mu}_\alpha^{\rm eq} - (\vec{\bar{\mu}}^\alpha)^T[\boldsymbol{\chi}^\alpha]\vec{\mu}_\alpha^{\rm eq} - (\Delta\vec{C}_{\rm eq}^\alpha)^T(\vec{\mu} - \vec{\mu}_\alpha^{\rm eq})
\end{aligned} \tag{9.93}$$

Substituting Eq. (9.91) and Eq. (9.92) into Eq. (9.90) and making use of Eq. (9.93), finally gives, after tedious term-collecting,

$$\begin{aligned}
\omega^L(\vec{\mu}) - \omega^\alpha(\vec{\mu}) = &-\frac{1}{2}(\vec{\mu} - \vec{\mu}_\alpha^{\rm eq})^T\left([\boldsymbol{\chi}^L] - [\boldsymbol{\chi}^\alpha]\right)(\vec{\mu} - \vec{\mu}_\alpha^{\rm eq}) \\
&- (\Delta\vec{C}_{\rm eq}^\alpha)^T(\vec{\mu} - \vec{\mu}_\alpha^{\rm eq})
\end{aligned} \tag{9.94}$$

9.3.6 Casting the Driving Force in Terms of Supersaturation

It is instructive to re-cast Eq. (9.94) in a form that is more easily comparable to the form of the binary model in the previous section. We define a reduced supersaturation vector $\vec{U}_\alpha$ associated with each solid phase (α) by

$$\vec{U}_\alpha = \frac{[\boldsymbol{\chi}^L]}{|\Delta\vec{C}_{\rm eq}^\alpha|}(\vec{\mu} - \vec{\mu}_\alpha^{\rm eq}) \tag{9.95}$$

whose components represent, as before, the local difference in each component's chemical potential from its corresponding [local] equilibrium value. We also define a corresponding *concentration normal vector* $\hat{n}_c$ by

$$\hat{n}_c = \frac{\Delta \vec{C}^{\alpha}_{\text{eq}}}{|\Delta \vec{C}^{\alpha}_{\text{eq}}|}, \tag{9.96}$$

where $||$ denotes the usual vector norm (and $|\hat{n}_c| = 1$). In terms of Eq. (9.95) and Eq. (9.96), Eq. (9.94) can be written in the compact form

$$\omega^{\alpha}(\vec{\mu}) - \omega^{L}(\vec{\mu}) = |\Delta \vec{C}^{\alpha}_{\text{eq}}|^2 \left\{ \vec{U}^T_{\alpha} \frac{(\boldsymbol{I} - [\boldsymbol{K}^{\alpha}])}{2} + \hat{n}^T_c \right\} [\boldsymbol{\chi}^L]^{-1} \vec{U}_{\alpha} \tag{9.97}$$

Equation (9.97) is exactly the matrix analogue of the driving force used in the binary alloy phase field equations in previous sections, except that here it is applicable to multiple components.

9.3.7 Final Form of Phase Field Equations in Terms of Supersaturation Driving Forces

We summarize here the final form of the the phase field equations in terms of the above definitions. The dynamical evolution equations for each order parameter and for the chemical potential (corrected by anti-trapping matrix) become,

$$\begin{aligned} \tau_{\alpha} \frac{\partial \phi_{\alpha}}{\partial t} &= W^2_{\alpha} \boldsymbol{\nabla}^2 \phi_{\alpha} - f'_{\text{DW}}(\phi_{\alpha}) - w_{obs} \phi_{\alpha} \sum_{\beta \neq \alpha}^{N} \phi^2_{\beta} \\ &- \left(\frac{(\boldsymbol{I} - [\boldsymbol{K}^{\alpha}])^T}{2} \vec{U}_{\alpha} + \hat{n}_c \right)^T [\lambda_{\alpha}] \vec{U}_{\alpha} \, g'_{\alpha}(\boldsymbol{\phi}) \end{aligned} \tag{9.98}$$

and

$$\begin{aligned} [\chi] \frac{\partial \vec{\mu}}{\partial t} &= \boldsymbol{\nabla} \cdot \left[[\boldsymbol{M}] \boldsymbol{\nabla} \vec{\mu} + \sum_{\alpha} W_{\alpha} [a(\boldsymbol{\phi})] \, |\Delta \vec{C}^{\alpha}_{\text{eq}}| \left\{ \hat{n}_c + (\boldsymbol{I} - [\boldsymbol{K}^{\alpha}])^T \, \vec{U}_{\alpha} \right\} \frac{\partial \phi_{\alpha}}{\partial t} \frac{\boldsymbol{\nabla} \phi_{\alpha}}{|\boldsymbol{\nabla} \phi_{\alpha}|} \right] \\ &+ \frac{1}{2} \sum_{\alpha} |\Delta \vec{C}^{\alpha}_{\text{eq}}| \left\{ \hat{n}_c + (\boldsymbol{I} - [\boldsymbol{K}^{\alpha}])^T \, \vec{U}_{\alpha} \right\} \frac{\partial \phi_{\alpha}}{\partial t}, \end{aligned} \tag{9.99}$$

where the all the variables in these equations are summarized as follows:

$$\vec{U}_{\alpha} = \frac{[\boldsymbol{\chi}^L]}{|\Delta \vec{C}^{\alpha}_{\text{eq}}|} (\vec{\mu} - \vec{\mu}^{\text{eq}}_{\alpha}) \tag{9.100}$$

$$\hat{n}_c = \frac{\Delta \vec{C}^{\alpha}_{\text{eq}}}{|\Delta \vec{C}^{\alpha}_{\text{eq}}|} \tag{9.101}$$

$$[\boldsymbol{K}^\alpha] = [\boldsymbol{\chi}^L]^{-1}\,[\boldsymbol{\chi}^\alpha] \tag{9.102}$$

$$[\lambda_\alpha] = \hat{\lambda}_\alpha |\Delta\vec{C}^\alpha_{\rm eq}|^2 [\chi^L]^{-1} \tag{9.103}$$

$$\Delta\vec{C}^\alpha_{\rm eq} = \vec{c}^{\,L}_{\rm eq} - \vec{c}^{\,\alpha}_{\rm eq} \tag{9.104}$$

$$[\boldsymbol{\chi}] = [\boldsymbol{\chi}^L]\left\{\mathbf{I} - \sum_\alpha \left(\mathbf{I} - [\boldsymbol{K}^\alpha]\right) h_\alpha(\vec{\phi})\right\} \tag{9.105}$$

$$[\boldsymbol{M}] = \sum_\alpha^N q_\alpha(\phi)[\boldsymbol{D}]^\alpha\,[\boldsymbol{\chi}^\alpha] + \left(1 - \sum_\alpha^N q_\alpha(\phi)\right)[\boldsymbol{D}^L]\,[\boldsymbol{\chi}^L] \tag{9.106}$$

The outer gradient operator in Eq. (9.99) contracts with the index of the inner gradients for the first and second terms on the right-hand side of Eq. (9.99), as in Eq. (6.6). The matrices $[\boldsymbol{\chi}^L]$ and $[\boldsymbol{\chi}^\alpha]$ are given by Eq. (9.62). The parameters τ_α, W_α, $\hat{\lambda}_\alpha$, w_{obs} and interpolation functions are as defined in the previous sections. It is noted that for this general form of multi-component alloys, the ant-trapping must be represented as a *matrix* of anti-trapping fluxes. It is clear that equations Eq. (9.98) and Eq. (9.99) are the multi-component (vector) analogues of Eq. (9.56) and Eq. (9.52)[9], and in the special case where all matrices become diagonal, these equations collapse back to equations Eq. (9.16) and Eq. (9.19).

In the above model, the susceptibility matrix $[\boldsymbol{\chi}]$ and mobility matrix $[\boldsymbol{M}]$ need to be evaluated at each point in space, while $[\boldsymbol{\chi}^L]^{-1}$, $[\boldsymbol{\chi}^\alpha]$ and $[\boldsymbol{K}^\alpha]$ need only be computed once (for a given temperature). It is noted that for a more robust description of realistic alloys, the coefficient matrices in the equations of motion can be made $\vec{\mu}$ (or $\vec{c}$) dependent, thus adopting changing values that can better capture the properties of liquid and solid over different broad ranges of concentration space that can occur in the system. This is a complex operation requiring significant off-line work with materials databases to create and store arrays of these properties at different temperature and concentration ranges. Nevertheless, it is in theory possible if accuracy is desired. However, for most studies involving the physics of solidification, using constant coefficient matrices in the above model is usually sufficient.

The asymptotic analysis of Eq. (9.98) and Eq. (9.99) goes beyond the scope of the analysis done in Appendix (B), which is applicable to binary alloys or multi-component alloys with diagonal property matrices. One practical approach is that used to analyze Eqs. (9.16) and (9.19) in Section 9.1.3, and base the choice of parameters of this more complex

[9]Note, this equivalence is formally true before the re-shaping of $q(\phi) \equiv M/D_L$ in Eq. (9.52) into $q(\phi) = \tilde{q}(\phi)\chi(\phi)$ as was done in Eq. (9.54).

model on the diagonal components, or even the component with the slowest mobility. One can also complement this by a numerical analysis to determine the amplitude of the coefficients of the anti-trapping matrix, assuming, for simplicity, that each component takes on the same functional form.

CHAPTER 10

Application: Phase Field Modelling of Ternary Alloys

It is instructive at this juncture to demonstrate a concrete application of the grand potential phase field modelling formalism developed in previous chapters. This chapter specializes the phase field model formulation of Section 9.3 to the tungsten-carbon-cobalt (W-C-Co) ternary system. Here, rather than apply the phase field equations to usual dendrite morphology during solificiation growth, we opt here to demonstrate the model's applicability to the dissolution of tungsten carbide (WC) in a cobalt matrix during thermal spray coating deposition.

10.1 THERMAL SPRAY COATING DEPOSITION OF WC-Co

Tungsten carbide-cobalt (WC-Co) is a metal matrix composite consisting of hard tungsten carbide (WC) particles and a softer Fcc cobalt matrix. It is commonly applied as a thermal spray coating to improve the wear resistance of a surface in harsh conditions [73]. In thermal spray coatings, the idea is to feed the material as as a composite WC-Co powder into a high velocity high temperature gas stream. The gas stream melts the cobalt matrix inside each powder particle, and propels the powder onto a substrate. Upon impact, the powder particles deform into high aspect ratio "splats", and accumulate into a lamellar coating with a thickness of several hundreds of micrometers [74, 75].

DOI: 10.1201/9781003204312-10

A common problem with the thermal spray deposition of WC-Co is that the excessive heating of droplets starts to dissolve the WC carbides into the cobalt melt. This dissolution embrittles the cobalt matrix via formation of secondary hard phases such as W_2C carbides and Co-W-C ternary carbides, as well as formation of nanocrystalline or amorphous cobalt-rich phase [73]. The cobalt matrix embrittlement decreases the wear resistance provided by the coating. On the other hand, if too little heat is applied to the powder particles, the cobalt matrix does not melt sufficiently, leading to imperfect contact between the WC phase and the Fcc cobalt matrix. This decreases the ability of the hard carbide particles to strengthen the soft cobalt matrix, thereby decreasing the coating wear resistance. Therefore, there is a trade off between insufficient and excessive heating, which can be investigated with phase field modeling, in order to provide an optimal level of carbide dissolution in the coating.

10.2 REPRESENTATION OF THERMODYNAMIC PHASES

To proceed to investigate this phenomenon with phase field modeling, the relevant thermodynamic phases need to be identified: liquid, stoichiometric tungsten carbide (WC), and a cobalt-rich Fcc. The raw free energy data is gathered from the Thermo-Calc TCFE9 database, and for the two phases with measurable solubility (liquid and Fcc), the free energy curvatures are directly evaluated to approximate the free energies parabolically.

WC phase has no measurable solubility in cobalt, i.e. it is stoichiometric. In the phase field formalism all fields need to vary smoothly between phases, and therefore, a finite solubility is needed for each phase. For too small a solubility (too large a free energy curvature), even the slightest variation from the stoichiometry leads to an abrupt drop of the order parameter at an interface with liquid or another solid phase. Here, following the idea in Ref. [76], the stoichiometry is emulated with a sharp parabolic free energy, where the parabolic (free energy) curvature is assigned a large value, several times larger than phases with real solubility. This leads to a finite but small solubility, so that the concentration in these phases are sufficiently close to the intended stoichiometry.

With the above considerations in mind, the free energies of the phases of this system are expressed in the general parabolic form

$$G^{\vartheta}(\vec{c}) = \frac{1}{2}(\vec{c} - \vec{c}_{\vartheta}^{\,\min})^T \left[\mathbf{G}_{cc}^{\vartheta}\right] (\vec{c} - \vec{c}_{\vartheta}^{\,\min}) + (\vec{c} - \vec{c}_{\vartheta}^{\,\min})^T \vec{G}_c^{\vartheta} + G_{\vartheta}^{\min}, \quad (10.1)$$

where it is noted that the symbol G^ϑ is used instead of f^ϑ to denoted the free energy in this chapter. Here, $\vec{c}_\vartheta^{\,\text{min}}$ is the point around which the parabolic curvatures $\left[\mathbf{G}_{cc}^\vartheta\right]$ and linear terms $\vec{G}_c^\vartheta$ are pre-computed.

In what follows, the parabolas are assumed to have no cross-species terms, making the free energy curvature matrix diagonal,

$$\left[\mathbf{G}_{cc}^\vartheta\right] = \begin{bmatrix} G_{11}^\vartheta & 0 \\ 0 & G_{22}^\vartheta \end{bmatrix} \tag{10.2}$$

This makes the generalized susceptibility matrix diagonal as well,

$$\left[\boldsymbol{\chi}^\vartheta\right] := \left[\mathbf{G}_{cc}^\vartheta\right]^{-1} = \begin{bmatrix} 1/G_{11}^\vartheta & 0 \\ 0 & 1/G_{22}^\vartheta \end{bmatrix} \tag{10.3}$$

With these definitions, the parabolic phase free energy can be represented as a sum according to

$$G^\vartheta(\vec{c},T) = G_\vartheta^{\text{min}} + \sum_{j=1}^{2}\left[\frac{1}{2}G_{jj}^\vartheta(c_j - c_{\vartheta j}^{\text{min}})^2 + G_j^\vartheta(c_j - c_{\vartheta j}^{\text{min}})\right], \tag{10.4}$$

where G_ϑ^{min}, G_j^ϑ, and G_{jj}^ϑ are constructed from the full CALPHAD free energies at several temperatures, and in the simulations these values are interpolated with respect to temperature.

10.3 TABULATED TIELINES IN THE LOW SUPERSATURATION LIMIT

In the low supersaturation limit, for each solid phase α, the equilibrium or metastable tielines with respect to the liquid phase need to be predetermined. Therefore, we need a mapping for a given input concentration $\vec{c}_{\text{input}}$ and temperature T to a tieline:

$$(\vec{c}_{\text{input}}, T) \rightarrow \vec{c}_{\text{eq}}^{\,\alpha}, \vec{c}_{\text{eq}}^{\,L}, \vec{\mu}_{\text{eq}} \tag{10.5}$$

where $\vec{c}_{\text{eq}}^{\,\alpha}$ and $\vec{c}_{\text{eq}}^{\,L}$ are the tieline compositions on the solid (α) side and liquid (L) side, respectively, and where each of $\vec{c}_{\text{eq}}^{\,\alpha}$ and $\vec{c}_{\text{eq}}^{\,L}$ have two components, which we'll denote by $c_{\text{eq}\,1}^\vartheta$ and $c_{\text{eq}\,2}^\vartheta$, $\vartheta = \alpha$, L. As well, we define for each temperature and equilibrium chemical potential on that tieline $\vec{\mu}_{\text{eq}} = \vec{G}_{\text{eq}\,c}$. For the W-C-Co ternary system we consider, the average concentration reflects the average composition of the system, which is 12 wt% WC and 88 wt% Co.

In principle, the mapping of Eq. (10.5) needs to be determined throughout the coexistence region of the system's phase diagram, and also in the metastable compositions and temperatures, where the solid α is metastable or the local composition $\vec{c}$ is outside the coexistence region. The tieline mapping of Eq. (10.5) is evaluated for a range of concentrations covering $\vec{0} \leq \vec{c}_{\rm input} \leq 1$ ($0 \leq c_{\rm input\,1} \leq 1$ and $0 \leq c_{\rm input\,2} \leq 1$) at various temperatures, and stored in a multidimensional tieline table. The tielines are constructed by enforcing equilibrium conditions, which enforce, first, the equality of chemical potentials for each component labeled 1 and 2, i.e.,

$$\mu_{\rm eq\,1} = \mu^{\alpha}_{\rm eq\,1} = \mu^{L}_{\rm eq\,1}, \tag{10.6}$$

$$\mu_{\rm eq\,2} = \mu^{\alpha}_{\rm eq\,2} = \mu^{L}_{\rm eq\,2}, \tag{10.7}$$

and, second, the equality of grand potentials, i.e. pressures,

$$G^{\alpha} - c^{\alpha}_{\rm eq\,1}\,\mu_{\rm eq\,1} - c^{\alpha}_{\rm eq\,2}\,\mu_{\rm eq\,2} = G^{L} - c^{L}_{\rm eq\,1}\,\mu_{\rm eq\,1} - c^{L}_{\rm eq\,2}\,\mu_{\rm eq\,2} \tag{10.8}$$

Moreover, we demand that each tieline, of the form $c_2^{\vartheta} = K c_1^{\vartheta} + B$ in c_1-c_2 space, intersects the input (i.e., local "average") concentrations $\vec{c}_{\rm input} = (c_{\rm input\,1}, c_{\rm input\,2})$, and the equilibrium concentration of the solid, $\vec{c}^{\alpha}_{eq}$, and the liquid phase $\vec{c}^{L}_{eq}$ for some slope K and off-set B, i.e.,

$$c^{\alpha}_{\rm eq\,2} = K c^{\alpha}_{\rm eq\,1} + B \tag{10.9}$$

$$c^{L}_{\rm eq\,2} = K c^{L}_{\rm eq\,1} + B \tag{10.10}$$

The six equations above can be used to determine the six unknowns: equilirium concentrations of solid α and liquid ($c^{\alpha}_{\rm eq\,1}$, $c^{\alpha}_{\rm eq\,2}$; $c^{L}_{\rm eq\,1}$, $c^{L}_{\rm eq\,2}$), and the tieline slope K and off-set B.

An example of the isothermal ternary phase diagram constructed from the above procedure is shown in Figure 10.1 at 1585 K, showing the exact (full) thermodynamic data on the left and parabolic approximation on the right, where WC has a finite solubility (light blue region in the top right corner). Increasing the WC free energy curvature shrinks the light blue region, and it reduces to a stoichiometric point in the limit of infinite free energy curvature.

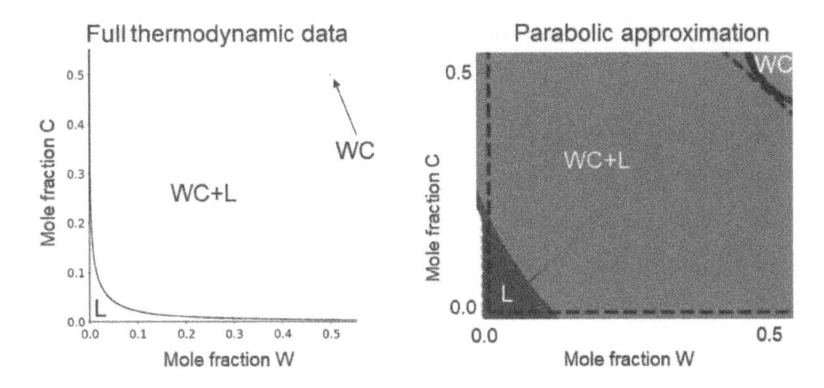

Figure 10.1 Phase diagrams for a ternary W-C-Co system with WC and liquid phases at 1585 K. Left: full thermodynamic data including a point-like stoichiometric WC phase in the upper right corner. Right: parabolic free energy fits for both liquid (L) and WC, where the light grey in the top right corner region shows the numerical solubility range of WC and the darker grey in the bottom left the shows the solubility of the liquid. (Solubility regions shown in the colour eBook.)

10.4 MOBILITY AND DIFFUSION COEFFICIENTS IN TERNARY SYSTEMS

As we have seen previously, the diffusion of species i can be expressed as

$$\frac{\partial c_i}{\partial t} = \nabla \cdot \boldsymbol{J}_i \tag{10.11}$$

where the flux species i is

$$\boldsymbol{J}_i = -\sum_{j=1}^{K} M_{ij} \nabla \mu_j(\boldsymbol{c}) \tag{10.12}$$

To use the formalism derived previously in this book, at this juncture we require a concrete description of the mobility matrix or conversely the diffusion matrix. If the mobility data is available for the material, for example in software databases, such as in DICTRA, then the mobilities M_{ij} can be evaluated and used directly. However if only the diffusion coefficients are available, they can be used to construct the equivalent mobility using the chemical potential gradient representation of Eq. (10.12) and a matching with a Fickian flux expression through a chain rule as follows,

$$\boldsymbol{J}_i = -\sum_{j=1}^{K} M_{ij}\left[\sum_{k=1}^{K} \frac{\partial \mu_j}{\partial c_k}\nabla c_k\right] = -\sum_{k=1}^{K}\left(\sum_{j=1}^{K} M_{ij}\frac{\partial \mu_j}{\partial c_k}\right)\nabla c_k = -\sum_{k=1}^{K} D_{ik}\nabla c_k, \tag{10.13}$$

where we have identified the relationship between diffusion and mobility using the the definition

$$D_{ik} = \sum_{j=1}^{K} M_{ij}\frac{\partial \mu_j}{\partial c_k} \tag{10.14}$$

(in these sums, K index components). Denoting $\partial\mu_j/\partial c_k$ by G_{jk} (second partial derivative of the phase-specific free energy with respect to components j and k), we obtain the following system of equations for the diffusion matrix of a ternary alloy,

$$\begin{bmatrix} D_{11} & D_{12} \\ D_{21} & D_{22} \end{bmatrix} = \begin{bmatrix} M_{11} & M_{12} \\ M_{12} & M_{22} \end{bmatrix} \cdot \begin{bmatrix} G_{11} & G_{12} \\ G_{12} & G_{22} \end{bmatrix}$$

$$\begin{bmatrix} M_{11} & M_{12} \\ M_{12} & M_{22} \end{bmatrix} = \frac{1}{G_{11}G_{22} - G_{12}^2}\begin{bmatrix} D_{11}G_{22} - D_{12}G_{12} & D_{12}G_{12} - D_{11}G_{12} \\ D_{21}G_{22} - D_{22}G_{12} & -D_{21}G_{12} + D_{22}G_{11} \end{bmatrix} \tag{10.15}$$

Assuming no cross-diffusion, i.e. $D_{12}, D_{21} \approx 0$, the mobility matrix thus reduces to

$$\mathbf{M}^{\vartheta} = \frac{1}{G_{11}G_{22} - G_{12}^2}\begin{bmatrix} D_{11}G_{22} & -D_{11}G_{12} \\ -D_{22}G_{12} & D_{22}G_{11} \end{bmatrix}, \tag{10.16}$$

which demonstrates that the mobility matrix still contains off-diagonal terms, and that a diagonal diffusion matrix can still cause off-diagonal fluxes driven by chemical potential gradients. Only in the limit $G_{12} \ll G_{11}, G_{22}$ does this reduce to a diagonal mobility matrix,

$$\mathbf{M}^{\vartheta} = \begin{bmatrix} D_{11}/G_{11} & 0 \\ 0 & D_{22}/G_{22} \end{bmatrix}. \tag{10.17}$$

The W and C diffusivities in liquid are taken from electronic structure calculations [77], with an Arrhenius temperature dependence: $D(T) = D_0 \cdot \exp(-Q/(RT))$, where D_0 is a constant prefactor and Q is the activiation energy. Solid diffusivity is assumed to be zero in the remainder of this chapter.

10.5 LOW SUPERSATURATION LIMIT OF A TERNARY ALLOY IN THE GRAND POTENTIAL PHASE FIELD MODEL

The preceding considerations allow us to specialize the order parameter equations of Section 9.3 to

$$\tau_\alpha \frac{\partial \phi_\alpha}{\partial t} = W_\alpha^2 \nabla^2 \phi_\alpha - f'_{\mathrm{DW}}(\phi_\alpha) - \omega_{obs}\phi_\alpha \sum_{\beta\neq\alpha} \phi_\beta^2 - \sum_j \lambda_\alpha^j U_{\alpha j} g'(\phi_\alpha),$$

for $\alpha =$ Fcc and WC, (10.18)

while the dynamical equation for the chemical potentials becomes

$$X_{jj}\frac{\partial \mu_j}{\partial t} = \boldsymbol{\nabla}\cdot\left[M_{jj}\boldsymbol{\nabla}\mu_j + \sum_\alpha W_\alpha a_t(\phi_\alpha)|\Delta C_j^\alpha|\left(\hat{n}_{c\,j} + (1-K_j^\alpha)U_{\alpha\,j}\right)\frac{\partial \phi_\alpha}{\partial t}\frac{\nabla\phi_\alpha}{|\nabla\phi_\alpha|}\right]$$
$$+\frac{1}{2}\sum_\alpha |\Delta C^\alpha_{\mathrm{eq}\,j}|\left(\hat{n}_{c\,j} + (1-K_j^\alpha)U_{\alpha\,j}\right)\frac{\partial\phi_\alpha}{\partial t}, \quad \text{for } j = 1,2, \tag{10.19}$$

where in Eq. (10.19) the phase-interpolated generalized susceptibility becomes

$$X_{jj} = \sum_\alpha g(\phi_\alpha)\frac{1}{G^\alpha_{jj}} + \left(1 - \sum_\alpha g(\phi_\alpha)\right)\frac{1}{G^L_{jj}}, \tag{10.20}$$

and the diagonal mobility components are given by

$$M_{jj} = \left(1 - \sum_\alpha g(\phi_\alpha)\right)\frac{D^L}{G^L_{jj}}. \tag{10.21}$$

Following the definition in Eq. (9.82), the generalized partition coefficient for each solute component (comparable to the partition coefficient in dilute binary alloys) becomes

$$K_j^\alpha = \frac{G^\alpha_{jj}}{G^L_{jj}} \tag{10.22}$$

The concentration difference between the two ends of a Liquid-α tieline is given in component form by

$$\Delta C^\alpha_{\mathrm{eq}\,j} = c^L_{\mathrm{eq}\,j} - c^\alpha_{\mathrm{eq}\,j}, \tag{10.23}$$

while the norm of concentration difference is given by

$$\hat{n}_{c\,j} = \frac{\Delta C^\alpha_{\mathrm{eq}\,j}}{|\Delta C^\alpha_{\mathrm{eq}\,j}|} \tag{10.24}$$

Finally, in terms of the above definitions, the supersaturation fields for this ternary system are given in component form by

$$U_{\alpha j} = \frac{1}{|\Delta C_j^\alpha| G_{jj}^L} (\mu_j - \mu_{\text{eq}\, j}) \tag{10.25}$$

10.6 PHASE FIELD PARAMETERS FOR EMULATING THE SHARP INTERFACE LIMIT

To set the phase field model parameters we follow the procedure outlined in Section 9.1.6. We can define a solutal capillary length of species j by

$$d_{oj} = \frac{2\sigma_{SL}}{(\Delta C_{\text{eq}\, j}^\alpha)^2 G_{jj}^L}, \tag{10.26}$$

where $\Delta C_{\text{eq}\, j}^\alpha = c_{\text{eq}\, j}^L - c_{\text{eq}\, j}^\alpha$ is the concentration difference between solid α and liquid L, and G_{jj}^L is free energy curvature of the liquid phase, respectively, as defined above. By analogy with the first line of Eq. (9.36), the reference coupling constant is then given by

$$\lambda_\alpha^j = a_1 \frac{W_\alpha}{d_{o\, j}}. \tag{10.27}$$

It is noted that if we scale out one of the coupling constants λ_α^m (m is one of the compositions) in Eq. (10.18), we would arrive back to the form of the driving force for the α order parameters that looks like Eq. (9.25), assuming a solid-liquid surface energy that is independent of composition. Moreover, doing so would allow the Gibbs-Thomson interface condition to be written in the form of Eq. (9.31), with one capillary length of the form in Eq. (9.35). In the results presented below, we match the capillary length using a solid-liquid interface energy estimated by Jian et al. for pure Cobalt at 300 K below the melting point, i.e. $\sigma_{SL} = 0.235$ J/m^2 [78].

The second order thin-interface analysis for the general case of the model in Section 9.3 is not presently available. For the case of diagonal property matrices that we are considering in this section, we can use the relationships in Eq. (9.36). However, since the example shown below will concern dissolution of carbides into the Co matrix, the effect of the kinetic coefficient inherent in the phase field equations Eq. (10.18) and Eq. (10.19) has a negligible effect on the kinetics at the interface. For simplicity, therefore, we chose the solute component m that gave the

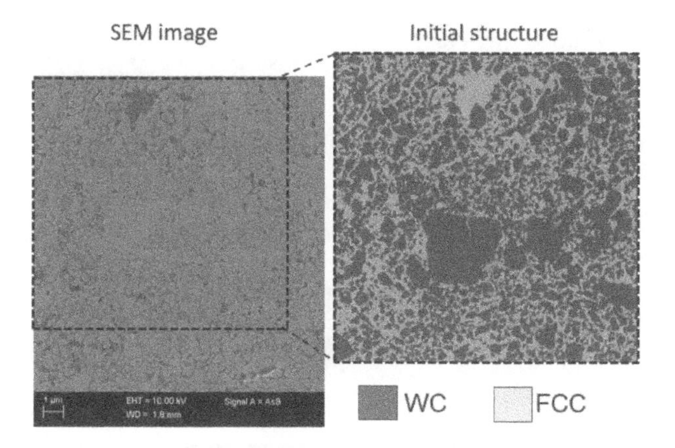

Figure 10.2 Segmentation of an scanning electron microscope image of WC-Co microstructure (left) into an initial condition for phase field simulations (right). The grey scale in the right image represents the different solid order parameters. (Different order parameters are shown in the colour eBook.)

smallest time scale τ_α, as determined by the second line of Eq. (9.36), namely

$$\tau_\alpha^m = \frac{\lambda_m^\alpha W_\alpha}{a_1}\left(\beta + a_1 a_2 \frac{W_\alpha}{D_m^L}\right) \tag{10.28}$$

where D_m^L is the liquid-side solute diffusion coefficient of species m, which we take here as carbon. A more precise value for β is given by a sum-like expression in Eq. (9.36). However, here β is assumed to be zero and its effect is negligible as we are concerned with dissolution of the carbides.

10.7 SIMULATIONS OF CARBIDE DISSOLUTION

This section shows the results of simulations of the ternary grand potential phase field model presented in the previous section that demonstrate the dissolution of tungsten-carbine into a cobalt matrix. To initialize the WC-Co metal-matrix composite microstructure, we used a scanning electron microscope (SEM) image, which was segmented into an initial distribution of WC and Fcc cobalt, as shown in Figure 10.2.

From thermal spray process simulations [79, 80] it is found that the range of typical powder particle temperatures during deposition are estimated to vary between 1250 K and 1750 K and the particle flight times

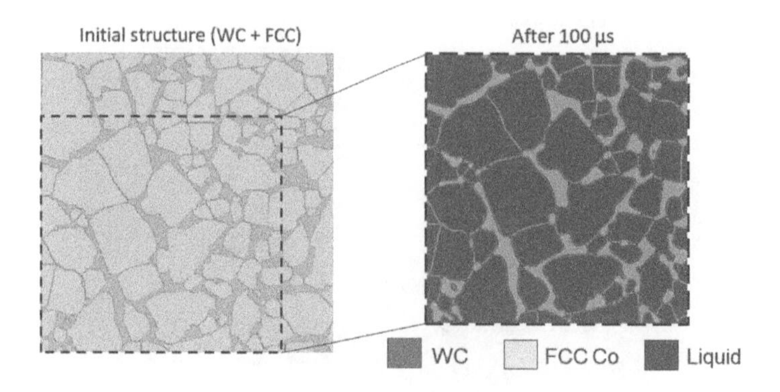

Figure 10.3 Synthetic WC-Co microstucture (left) input into simulations at 1500 K, showing minor dissolution after $100\mu s$ and the emergence of a very small liquid channels appearing between order parameters of the WC and Fcc Co phases, which are represented by dark and light grey scales, respectively. (Shown by dark and light colours in the eBook.)

are on the order of milliseconds. The exact temperature history of the particles varies significantly, depending on the the powder size and exact trajectory of the powder particle in the turbulent gas stream. Here, we assume that the powder particle temperature is constant, which is a reasonable approximation for relatively large powder particles, such as those in the size range of 10–40 μm.

An example simulation demonstrating the dissolution of Cobalt is shown in Figure 10.3 for a synthetically packed WC-Co microstructure. The right frame shows the phase distribution after 100 μs at 1500 K, showing minor dissolution and melting of the order parameters corresponding to the cobalt phase.

Figure 10.4 shows the dissolution at three different temperatures (1250 K, 1500 K, 1750 K) at early and late times, with each case starting from the same WC-Co microstructure shown in the left frame. At 1250 K, the original carbide structure is maintained without formation of a liquid phase and the carbide distribution remains unaltered. It is noted that while there is no harmful carbide dissolution, this microstructure is expected to manifest carbide-matrix detachment at this processing temperature due to the rapid deformation of the powder particle during its impact to the substrate (or a previous powder layer).

For the case of dissolution at 1750 K shown in Figure 10.4, the Fcc cobalt becomes fully liquid, and the smaller WC domains start to

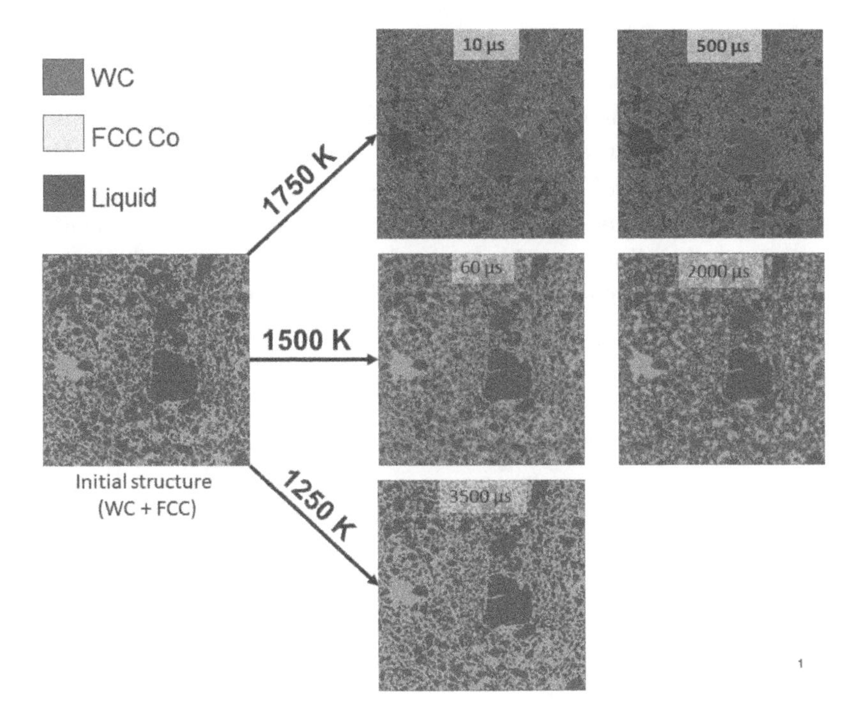

Figure 10.4 Phase field model simulations of carbide dissolution at three deposition temperatures and for two dwell times. The left frame shows the image-segmented WC-Co initial microstructure, with lighter grey repressing FCC Co and darker grey representing WC. In the right-hand frames for T=1500 K and 1250 K, darker grey colours represent WC and lighter colours FCC Co. In the T=1750 K images, dark grey colours represent either WC or liquid. (Shown in colour eBook, where for T=1750 K, red is WC and blue is liquid.)

dissolve and in some places merge together. The dissolution zones act as templates for subsequent formation of ternary Co-W-C carbides in the matrix, which leads to increased brittleness and the loss of toughness of the material. In addition, the aforementioned merging of carbide domains into a more contiguous "skeleton" reduces the ability of the metal-matrix composite system to transfer mechanical load smoothly between the matrix and the carbide phase. Therefore, particle flight at 1750 K is expected to result in excessively brittle coatings, thereby reducing the intended wear resistance properties of the coating.

Figure 10.4 also shows dissolution at 1500 K, which sits between the two extremes of 1250 K and 1750 K. Here, minor dissolution is seen, but the initial carbide distribution is well maintained. These results indicate that temperatures around approximately 1500 K define a suitable range for deposition of WC-Co thermal spray coatings. If larger carbide domains are engineered into the powder particles, higher deposition temperatures can be allowed and, conversely, lower deposition temperatures are suitable for samples containing smaller carbide domains.

III

CHAPTER 11

Interpreting Asymptotic Analyses of Phase Field Models

THIS CHAPTER DELVES A little deeper into the meaning of an *asymptotic analysis* of a phase field model (hereafter just called "asymptotics" for short). We will guide the discussion by using the backdrop of the single-component and single-order parameter limit of Eqs. (6.6) and (6.9). We discuss, in particular, how these equations converge onto the classic sharp interface model (SIM) of solidification described by Eqs. (9.29) and (9.30)[1] in the limit of low supersaturation, which we saw is also described by Eqs. (9.16) and (9.19). The details of the matched asymptotic analysis is covered in Appendix (B), and their practical application was discussed in earlier chapters. The main results we will reference here are given by Eqs. (B.71) and (B.84), along with Eq. (B.20) (with definitions of variables given therein). It is noteworthy that Eqs. (B.71) and (B.84) contain three "correction" terms, ΔF, ΔH and ΔJ that do not appear in the *classical* SIM. As such, these terms are unphysical and must be eliminated if we wish to emulate the *classical* SIM described by Eqs. (9.29) and (9.30) in the limit of $W_\phi \sim d_o$ (or

[1] It is noted that in general the constants a_1 and a_2 in Eqs. (9.29) and (9.30) come from integrals that depend on the lowest order steady state solutions of the phase field equations and the choice of interpolation functions used to modulate the grand potential density though the interface. Interpolation functions comprise degrees of freedom in phase field models that can be chosen for expediency. This is possible because the interface structure within the zone of the phase field interface is not represented quantitatively.

DOI: 10.1201/9781003204312-11

even for larger W_ϕ). It turns out that $\Delta H = 0$ and $\Delta J = 0$ for the interpolation functions in Eq. (6.12), while ΔF can be made zero by using the anti-trapping function and interpolation function for mass diffusion shown in Ref. [51]. These corrections will thus not be discussed further in this section and the reader is referred to Ref. [38] for the mathematical details. It is also noted that the SIM limit described by Eqs. (9.29) and (9.30) holds identically for each grain of the binary phase field model in Section 9.2.

11.1 WHAT IS AN ASYMPTOTIC ANALYSIS OF A PHASE FIELD MODEL ABOUT?

When there is scale separation between the interface width of the order parameter (W_ϕ) and the length scales that control the growth of microstructure (e.g. capillary length, diffusion length, radius of curvature), the kinetics of a phase field (PF) model typically emulate a corresponding *free-boundary problem*, described by a *sharp interface model (SIM)*. For low rates of solidification this is described by the *classical Stefan model*, which describes heat or solute diffusion in bulk phases accompanied by heat/solute flux conservation across the interface, and the classical Gibbs-Thomson curvature correction of the temperature/chemical potential. *Matched asymptotic boundary layer analysis* is the formal name of the mathematical analysis used to map the behaviour of a set of phase field equations to a corresponding SIM. The claim that a phase field model is *quantitative* simply means that model parameters and interpolation functions can be chosen such that its kinetics match those of a SIM in the limit of a diffuse (rather than microscopically sharp) interface width.

Phase field models of solidification have been known to formally reduce to the classic SIM at low solidification rates in the so-called *sharp-interface* limit where the interface width $W_\phi \rightarrow 0$ and the coupling constant $\lambda \rightarrow 0$, such that the ratio $W_\phi/\lambda \sim d_o$ (capillary length) and the phase field time scale satisfies $\tau/(W_\phi\lambda) \sim \beta \ll 1$ (kinetic coefficient) [60]. However, emulating SIM kinetics using these relationships requires microscopic values of W_ϕ and τ, making efficient numerical simulation effectively intractable due to the fine mesh and small time step resolutions required.

To remedy the numerical intractability of sharp-interface approaches, so-called *thin-interface* matched asymptotic analysis approaches for phase field modelling of solidification were developed [9, 10, 61] to emulate the classic SIM with *mesoscale* values of W_ϕ and τ. For alloys specifically, the analysis becomes complex due to the disparity in solid and liquid diffusion coefficients[2]. In this case, mapping an alloy onto the classic SIM with diffuse interfaces requires converting the phase field equations to a non-variational form as was done in Chapter 6, which involves: (1) treating the interpolation of diffusion and chemical potential through the interface as controllable degrees of freedom, and (2) adding the so-called, *anti-trapping flux* to the mass transport equation, as was introduced in Eq. (6.4). With these changes, thin-interface analyses can converge onto SIM results for a range interface widths, making it possible to conveniently increase the interface width and decrease numerical resolution.

11.2 UNDERSTANDING THE ROLE OF λ AS AN ASYMPTOTIC CONVERGENCE PARAMETER

When considering Eqs. (B.71) and (B.84), i.e. from which we extract the effective SIM coefficients for the capillary length (d_o) and kinetic coefficient (β), let us investigate further in what sense the phase field model "converges" to the classic SIM? After all, a phase field model has more parameters than the SIM. To answer this question, it is instructive to consider a phase field model (*Model C*) for a pure material, which can be easily shown to be a nearly identical constant-diffusion variant of the binary alloy grand potential phase field model. This is described by [38]

$$
\begin{aligned}
\tau\frac{\partial \phi}{\partial t} &= W_\phi^2 \nabla^2 \phi - g'(\phi) - \lambda \Delta \bar{T}\, U P'(\phi) \\
\frac{\partial U}{\partial t} &= D\nabla^2 U - \frac{1}{\Delta \bar{T}}\frac{\partial \phi}{\partial t},
\end{aligned}
\tag{11.1}
$$

where ϕ is the order parameter and $U = (T - T_m)/(T_\infty - T_m)$ is a dimensionless temperature field, with T_m being the melting temperature and T_∞ the far-field temperature of the melt. Here, D is used to denote the thermal diffusion coefficient, τ is the kinetic time scale of the order parameter equation, W_ϕ the interface width defined by the order parameter, $\lambda = L^2/(c_p T_m H)$ is the dimensionless coupling constant (where c_p and L are the specific heat and latent heat of fusion, respectively, and H is the nucleation barrier) and $\Delta \bar{T} = c_p(T_\infty - T_m)/L$ is the dimensionless far-field undercooling of the melt. For simplicity, we will

[2]It is noted that thermal diffusion in pure substances, denoted by α, is much larger than impurity diffusion. Typically, $D/\alpha \sim 10^{-3} - 10^{-4}$.

ignore anisotropy of the gradient terms as we will only be performing dimensional analysis of the above model in this section. The boundary conditions for this problem are $\phi = 0$ and $U \to 1$ at system boundaries, which are presumed to be very far from the growing crystal.[3]

Consider next a dendrite crystal with tip velocity V as illustrated in the cartoon in Figure 11.1. It is useful to analyze Eq. (11.1) in a co-

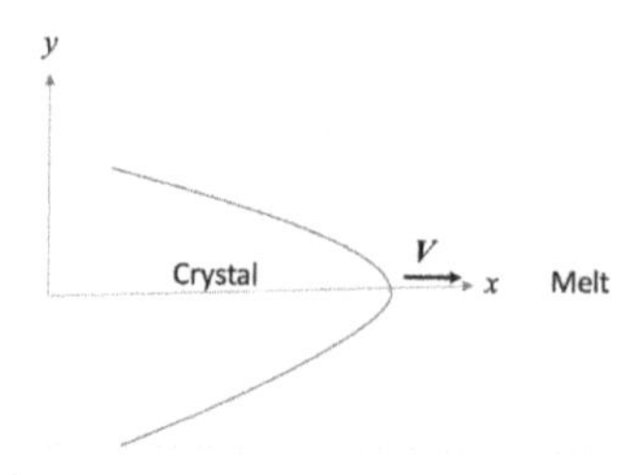

Figure 11.1 Schematic illustrating a crystal growing into an undercooled melt. The tip is assumed to be moving with a speed V.

ordinate system that is anchored to a position on the moving interface of the dendrite. Such a co-ordinate system is illustrated in Figure B.1 of Appendix (B.5) and is defined in terms of the variables (u, s), where u measures the transverse distance from the interface and s the arclength from some reference point on the interface. In these co-ordinates the Laplacian terms in Eq. (11.1) transform according to Eq. (B.21), while time derivatives transform according to Eq. (B.23). To simplify matters, we will analyze Eq. (11.1) along a line that passes directly through the tip. It will further be assumed that the dendrite tip has a large enough radius of curvature (or very small curvature) and that the solidification rate is slow enough that the fields do not vary explicitly with time near the tip. Under the first assumption, terms in Eq. (B.21) that depend on arclength s and κ can be dropped. The second assumption amounts to substituting the forms $\phi(x,t) = \phi(x - Vt)$ and $U(x,t) = U(x - Vt)$ into 1D versions of Eq. (11.1), which leads to the transformations $x \to u = x - Vt$, $\partial_x \to \partial_u$ and $\partial_t \to -V\partial_u$. Applying these transformations to

[3]While this situation describes solidification of a pure material to focus the discussion's essential features, the interpretations about the asymptotic convergence properties of a phase field model in the limit of a thin interface are generally valid for a binary alloys as well. The derivation of the above phase field model is covered in detail in Ref. [38].

Eq. (11.1) leads to

$$\frac{\partial^2 \phi}{\partial \xi^2} + \bar{V}\frac{d\phi}{d\xi} - g'(\phi) - \left(\lambda \Delta \bar{T}\right) U P'(\phi) = 0$$
$$\bar{D}\frac{\partial^2 U}{\partial \xi^2} + \bar{V}\frac{dU}{d\xi} + \left(\frac{\bar{V}}{\Delta \bar{T}}\right)\frac{d\phi}{d\xi} = 0, \tag{11.2}$$

where Eq. (11.2) has been made dimensionless by transforming space and time according to $\xi = u/W_\phi$ and $\bar{t} = t/\tau$, respectively, making $\bar{V} = \tau V/W_\phi$ the dimensionless tip velocity and $\bar{D} = D\tau/W_\phi^2$ the dimensionless diffusion coefficient.

It is clear that solutions of Eq. (11.2) can only depend on the parameters $(\bar{V}, \bar{D}, \lambda, \Delta\bar{T})$. In simulations of Eq. (11.1), it is observed that dendrites self-select a unique tip speed $\bar{V}$ for given values of the other parameters. Thus, Eq. (11.2) can only have consistent solutions for values of tip speed $\bar{V}$ that obey $\bar{V} = f_{\rm PF}(\bar{D}, \lambda, \Delta\bar{T})$, where $f_{\rm PF}$ is some unknown function. It is instructive to express τ and W_ϕ in terms of the material parameters of the SIM derived from the asymptotic analysis and listed in Eq. (9.30), namely, $d_o = a_1 W_\phi/\lambda$ and $\beta = a_1\tau/(W_\phi\lambda)\left\{1 - a_2\lambda/\bar{D}\right\}$. Since $\beta = 0$ in the *classic SIM*, this implies that $\bar{D} = a_2\lambda$, thus giving

$$\tau = \left(a_2/a_1^2\right) d_o^2 \lambda^3/D$$
$$W_\phi = d_o\lambda/a_1 \tag{11.3}$$

Equation (11.3) is then used to express the characteristic phase field model velocity $V_c^{\rm PF}$ as

$$V_c^{\rm PF} = W_\phi/\tau = \left(\frac{D}{d_o}\right)\frac{a_1}{a_2}\frac{1}{\lambda^2}, \tag{11.4}$$

in terms of which the aforementioned dimensionless tip velocity $\bar{V}$ can be written in terms of the dimensional velocity V as

$$\bar{V} = \frac{V}{V_c^{PF}} = \frac{V\tau}{W_\phi} = \left(\frac{Vd_o}{D}\right)\frac{a_2}{a_1}\lambda^2 \tag{11.5}$$

Along a separate line of inquiry, it is noted that the selection of dendrite tip speed in a pure material can also be predicted analytically by the so-called *microscopic theory of solvability (ms)* [81–87], which solves the classical SIM under conditions of a needle crystal growing in an undercooled melt. Microscopic solvabilty predicts that the steady state dendrite tip speed $V_{\rm ms}$ obeys $\bar{V}_{\rm ms} = V_{\rm ms} d_o/D = f_{\rm ms}(\Delta\bar{T})$ (it also couples to anisotropy but we are leaving that out of this discussion). In order for phase field simulations and *ms* predictions to agree. we must have $V_{\rm ms} d_o/D = V d_o/D$, which according to Eq. (11.5) implies

$$\frac{V_{\rm ms} d_o}{D} = \frac{V d_o}{D} = \frac{a_1}{a_2}\frac{\bar{V}}{\lambda^2} = f_{\rm ms}(\Delta\bar{T}) \tag{11.6}$$

Equation (11.6) and the scaling assumptions on tip speed made in the previous paragraph further implies that

$$f_{\rm PF}(\bar{D}, \lambda, \Delta\bar{T}) = f_{\rm PF}(a_2\lambda, \lambda, \Delta\bar{T}) \sim \lambda^2 f_{\rm ms}(\Delta\bar{T}) \tag{11.7}$$

The above considerations show that in order for the tip velocity predictions of the phase field model and microscopic solvability theory to be compatible, the former results must be divided by λ^2. This makes sense since λ relates to the nucleation barrier, a degree of freedom absent from the SIM, since the sharp interface theory only describes the kinetics of already moving interfaces.

In general, *quantitative* phase field simulations are performed in dimensionless units where space is scald by W_ϕ and time by τ, and by choosing a convenient value of λ, which acts as a convergence parameter proportional to the interface width (i.e., $\lambda \sim W_\phi/d_o$) and is used as an expansion parameter in the asymptotic analysis. Once obtained, the results of these simulations can then be scaled back to dimensional units by finding τ and W_ϕ using Eq. (11.3). These can be used to calculate the physical scale of spatial features in a simulation and the time of a field configuration, or to calculate the corresponding materials properties of the simulation (e.g. capillary length using $d_o = a_1 W_\phi/\lambda$, or diffusion coefficient using $D = (a_2/a_2^2)\, d_o^2 \lambda^3/\tau$). From there other properties can also be obtained, such as, for example, the local interface velocity (e.g. $V = (a_1/a_2)(D/d_o)(\bar{V}/\lambda^2)$, etc).

It is noted that the larger the value of λ used, the larger the possible system size that can be simulated and the longer the solidification times achieved. Of course, since λ is also proportional to the expansion parameter used to perform the asymptotic approximation of the fields, convergence of phase field simulations to physical SIM values will only be preserved if λ does not become too large. How large can λ be? Inspection of Eq. (11.2) shows that the solidification rate is proportional to the product $\lambda\Delta\bar{T}$, i.e. the product of the $\lambda\times$ thermodynamic driving force. Thus, when considering low solidification rates, it is possible to use values of $\lambda \sim \mathcal{O}(10^2)$ [10], while at very rapid solidification, $\lambda \sim \mathcal{O}(1)$ is suitable.

11.3 INTERPRETING THE ROLE OF λ IN ASYMPTOTIC ANALYSIS AND THE NOISE AMPLITUDE

As discussed above, in quantitative phase field simulations λ is treated as a "free" convergence parameter, typically chosen to smear the scale of the interface to make phase field simulations more numerically efficient. However, it is recalled that physically λ represents the inverse of the nucleation energy. It is also recalled from Chapter 8 that λ works its way into the amplitude of stochastic noise correlations in Eqs. (8.11) and (8.16) and thus plays a key role in nucleation process. It thus appears that the parameter λ may play two roles that are not fully consistent with each other.

To explore this further, consider noise (i.e., thermal fluctuations) only in the order parameter for the moment, although the discussion below is analogous for concentration or temperature noise. Equation (8.11) shows that noise strength of the order parameter field decreases as λ increases. This makes sense since increasing λ is the same as increasing W_ϕ in quantitative phase field simulations. As a result, the noise strength in a volume element that scales as $\sim W_\phi^d$ must decrease to match the noise strength corresponding to the physical volume that scales as $\sim d_o^d$, since $d_o \sim W_\phi/\lambda$.

When using noise in phase field models, fluctuations—like everything else—are only to be relevant on length scales larger than W_ϕ and time scales larger than τ. This is not a problem when realistic sizes of W_ϕ ($\sim$1 nm) are used as these are on the scale that can resolve a crystal nucleus. However, the volume of a crystal nucleus is typically much smaller than W_ϕ^d when $\lambda > 1$ is used, which implies that it is not possible to resolve realistic nuclei in the diffuse interface limit of phase field models. A plausible fix to this problem may be to reduce λ during the nucleation stage of a simulation, and then increasing λ later in the simulation when nuclei have grown sufficiently. Unfortunately, this will require an intractable number of numerical time steps until homogeneous nucleation actually occurs, since decreasing λ significantly reduces the time scale τ of each numerical time step (as well as requiring a more dense numerical mesh). This takes us back to the problem that necessitated quantitative phase field approaches and using large values of λ in the first place.

An alternate route is a compromise. Continue to use scaled up values of λ so as to facilitate numerical efficiency with which free boundary kinetics are simulated, in other words values or λ greater than one (e.g. it is typical to use $\lambda \sim \mathcal{O}(10 - 100)$ in low to moderate solidification rates). In this case, the phase field "nuclei" that emerge represent, very loosely speaking, a crystal "mass" that is actually the collective fusion of many microscopic nuclei formed over the coarse grained volume W_ϕ^d and over the coarse grained time scale τ–a type of "super-nucleus" so to speak. In other words, if we imagine coarse graining solidification microstructure on length and time scales much larger than W_ϕ and τ, respectively, such a "super-nucleus" can be seen as what emerges from the spatial and temporal coarse graining of the microscopic nucleation process. If, conversely, the length and time scales over which microstructure morphology is analyzed are on the order of W_ϕ and τ, respectively, there is no way to reconcile the results of matched asymptotic analysis and nucleation, i.e., it is not possible to simultaneously model nucleation and sharp-interface model kinetics quantitatively.

IV

VI

CHAPTER 12

The Regime of Rapid Solidification

TRADITIONAL CASTING PROCESSES OPERATE at low cooling rates, which are well described by the classic sharp interface model (*SIM*) (Eq. 9.29, with $\beta = 0$). The SIM assumes zero interface width, and that the interface is in near-equilibrium during solidification. This physically means that the solid-liquid interface is much smaller than the capillary length and that the interface moves slow enough that atomic attachment kinetics can maintain the same chemical potential on either side of the solid-liquid interface. Phase field modelling of solidification is often done in the binary alloy limit because it is simple in form and yet contains many physical processes found in more complex alloys. As we have discussed at length thus far, binary alloy phase field models (as well as some multi-component phase field models) can make quantitative predictions through the use of thin-interface approaches, at least in the limit of traditional casting phenomena where the interface moves slowly enough to be considered in local equilibrium.

In contrast, the equilibrium conditions that prevail in the classic SIM break down during rapid solidification. Here, the time scale of atomic attachment starts to approach the time scale for diffusion through the interface, invalidating the use of the classic SIM limit. Rapid solidification is becoming the norm in emerging technologies that collectively fit under the banner of "additive manufacturing" (AM). These include laser welding, 3D metal printing, thermal spray processes, and numerous others. AM is gaining traction due to the ability to rapidly produce complex net shape metal components with homogeneous microstructure and reduced waste. There is presently no one generally accepted SIM to describe the

DOI: 10.1201/9781003204312-12

rapid solidification processes relevant to these processes. Several sharp interface models have been proposed. The two most popular paradigms are the *continuous growth model* (CGM) of Aziz and co-workers [53, 54] and that of Sobolev and co-workers [88, 89]. The former works consider standard diffusion accompanied by attachment-limited kinetics at the interface, while the latter further incorporates two-time-scale dynamics to govern both inertial and diffusive dynamics of solute atoms near and through a rapidly advancing interface. While both modelling approaches fall under the banner of rapid solidification, the former is suitable to a slower range of solidification rates, while the latter is well suited to handle very high rates of solidification.

The works of Ahmad et al. [55], Wheeler et al. [90] and Boettinger et al. [91] showed that a phase field model of alloy solidification, governed by first order diffusion kinetics, captured most of the salient features of the CGM of Aziz and co-workers. However, these works also found that the fundamental CGM parameters extracted from any phase field model (e.g. the segregation coefficient $k(v_0)$ and kinetic undercooling) are sensitive to whatever phenomenological interpolation functions are designed into the phase field model to interpolate the free energy (or any other thermodynamic potential density) through the interface. Moreover, the connection between the phase field model and CGM parameters is non-trivial to derive mathematically, making it difficult to perform *quantitative* simulations of rapid solidification. This chapter will show how the results of the matched asymptotic analysis derived in the appendix can be manipulated to map the grand potential phase field model of a binary alloy onto the CGM model quantitatively.

CHAPTER 13

Modelling Continuous Growth Kinetics in the Diffuse Interface Limit of Grand Potential Phase Field Equations

THIS CHAPTER STUDIES THE non-equilibrium interface kinetic effects that emerge in the thin interface limit of Eqs. (6.6) and (6.9) for the case of a single crystal in a binary alloy. The aim is to derive an effective sharp interface model (SIM) from these equations that contains non-equilibrium corrections to the classic SIM, and compare this SIM with the continuous growth model (CGM) of solidification. We begin with a brief description of the continuous growth model of solidification. We then examine the results of Appendix (B), which performs a matched asymptotic analysis of Eqs. (6.6) and (6.9) to second order in the parameter $\epsilon = W_\phi/d_o$, i.e., the phase field interface relative to the capillary length. This analysis reveals that at non-zero interface velocities, the SIM derived from Eqs. (6.6) and (6.9) deviates from the classic SIM due to the presence of several new (often-called *"spurious"*) terms that arise due to a non-zero phase field interface width W_ϕ. Some of these spurious terms have been reported before for dilute, ideal alloys [10,51], while one more found here is new and arises due to the generality of the analysis presented here. Unlike the situation prevalent at low solidification rates,

DOI: 10.1201/9781003204312-13

as interface speed increases, these "spurious" terms cannot be neglected, even for small W_ϕ; indeed, they are physically relevant to any proper sharp interface description of rapid solidification at speeds of $\sim$ mm/s or larger. By considering the particular case of Henrian solutions, we show below that the asymptotic analysis of Appendix (B) maps the grand potential phase field equations onto the continuous growth model (GCM) of Aziz and co-workers [54].[1] Through the degrees of freedom afforded by the anti-trapping flux and the solute diffusion interpolation, it is shown how to control the solute segregation coefficient $k(v_0)$ and kinetic undercooling of the interface.

13.1 REVIEW OF THE CONTINUOUS GROWTH MODEL OF RAPID SOLIDIFICATION

As the solidification rate increases, the diffusion length becomes on the order of the solid-liquid interface, i.e. $D_L/v_0 \sim W_\phi$, where v_0 is the speed of the interface. In this limit, we expect two significant deviations from the classical sharp interface model in describing the physics of solidification. The first effect is that solute partitioning at the interface, defined by a *partition coefficient* $k \equiv c^\alpha/c^L$, becomes velocity dependent, $k \to k(v_0)$. This effect becomes significant when $v_0 \sim v_D \sim D_L/W_\phi$. For typical metals, $D_L \sim 10^{-9}$ m^2/s and $W_\phi \sim 10^{-9}$ m, making $v_D \sim 1$ m/s. It is known that $k(v_0) \to k_e$ (its equilibrium value) only for $v_0 \ll v_D$, making this kinetic effect important even for speeds somewhat lower than v_D.

Another effect that becomes important in rapid solidification is that the local composition and undercooling of the interface are dominated by the rate of attachment of atoms from the liquid to the solid, which sets the velocity scale of the advancing interface.[2] It has been proposed [53, 91, 92] that the solidification rate in this regime is described by

$$v_0 = v_c \left\{1 - \exp\left[\Delta G_{\text{eff}}/(RT/\Omega)\right]\right\}, \tag{13.1}$$

where ΔG_{eff} is the *effective free energy change per volume* of material solidified and v_c represents the rate of forward flux of atoms at the theoretical maximum driving force. Turnbull [92] considered $v_c \propto v_s$,

[1] The results of this section can also be applied to the single-crystal, binary alloy limit of Eqs. (9.16) and (9.19), which are derived from Eqs. (6.6) and (6.9) in the limit of low supersaturation.

[2] It is noteworthy that interface velocities in this regime are close to the absolute stability limit, making curvature effects of the interface negligible.

where v_s is the speed of sound of metal atoms in the liquid state, while Aziz and Boettinger [54] took $v_c = v_s$. Here we will also assume $v_c \approx v_s$.

The *effective* free energy density for solidification, ΔG_{eff} in Eq. (13.1), is typically written as in Ref. [93]

$$\Delta G_{\text{eff}} = \Delta G_{\text{sol}} - \Delta G_{\text{drag}} \tag{13.2}$$

The first term, ΔG_{sol}, is the thermodynamic driving force for solidification. It was shown by Cahn [94] to be

$$\Delta G_{\text{sol}} = f_\alpha(c^\alpha) - \left\{ f_L(c^L) - \left(c^L - c^\alpha\right) \frac{\partial f_L(c^L)}{\partial c} \right\} \tag{13.3}$$

where $c^\alpha(c^L)$ is the solid(liquid) concentration at the interface[3]. The effective driving force is reduced form the thermodynamic driving force by the so-called *drag force*, whose maximum value is given in Refs. [53–55, 93] as

$$\Delta G_{\text{drag}} = \left(c^L - c^\alpha\right)\left(\Delta\mu_A - \Delta\mu_B\right), \tag{13.4}$$

where $\Delta\mu_A(\Delta\mu_B)$ are the chemical potentials of the solvent(solute) in the solid minus that in the liquid, evaluated at the interface. The difference $\Delta\mu_A - \Delta\mu_B$ can be re-expressed in terms of the free energy function of each phase by using $\partial f_L(c^L)/\partial c = \mu_L^A - \mu_L^B$ and $\partial f_\alpha(c^\alpha)/\partial c = \mu_\alpha^A - \mu_\alpha^B$ [55].

Writing ΔG_{sol} and ΔG_{drag} in terms of $f_\alpha(c)$, $f_L(c)$ and their derivatives, recasts ΔG_{eff} in Eq. (13.2) as

$$\Delta G_{\text{eff}} = f_\alpha\left(c^\alpha\right) - f_L\left(c^L\right) + \left(c^L - c^\alpha\right) \frac{\partial f_\alpha(c_\alpha)}{\partial c} \tag{13.5}$$

For small velocities relative to the speed of sound, we expand the exponential in Eq. (13.1) as in Ref. [91]. Doing so yields

$$\Delta \bar{G}_{\text{eff}} = \bar{f}_\alpha\left(c^\alpha\right) - \bar{f}_L\left(c^L\right) + \left(c^L - c^\alpha\right) \frac{\partial \bar{f}_\alpha(c_\alpha)}{\partial c} = -\frac{v_0}{v_c}, \tag{13.6}$$

where over-bars imply energy densities have been re-scaled to be in units of RT/Ω.

[3]Ref. [94] assumes this driving force holds generally between any two phases undergoing a phase transformation.

13.1.1 Kinetic Undercooling of the Interface in Henrian Solutions

This section derives the kinetic undercooling at the interface implied by Eq. (13.6) for an alloy whose phases are described by Henry's law. The chemical part of the free energy of dilute liquid(solid) phase of such an alloy is described by

$$
\begin{aligned}
f_\alpha(c^\alpha,T) &= f_L(T_A) - \Delta T s_\alpha + \mathcal{E}_\alpha c^\alpha + \frac{RT_m}{\Omega}\left\{c^\alpha \ln c^\alpha - c^\alpha\right\} \\
f_L(c^L,T) &= f_L(T_A) - \Delta T\, s_L + \mathcal{E}_L c^L + \frac{RT_m}{\Omega}\left\{c^L \ln c^L - c^L\right\}
\end{aligned} \tag{13.7}
$$

where $\Delta T = T - T_A$, T_A is the melting temperature of component A, and $s_\alpha(s_L)$, $\mathcal{E}_\alpha(\mathcal{E}_L)$ are the entropy density and internal energy density of the solid(liquid), respectively.

Substituting Eq. (13.7), and its derivatives, into Eq. (13.5) gives

$$
\begin{aligned}
\Delta G_{\text{eff}} = &-\Delta T\,[s_\alpha - s_L] + [\mathcal{E}_\alpha - \mathcal{E}_L]\, c^L + (RT/\Omega) \\
&\times \left\{c^\alpha \ln c^\alpha - c^L \ln c^L + c^L - c^\alpha + \left(c^L - c^\alpha\right) \ln c^\alpha\right\}
\end{aligned} \tag{13.8}
$$

For dilute phases, we approximate $s_\alpha - s_L = -L_A/T_A = -(RT/\Omega)(1 - k_e)/|m_L^e|$, where L_A is the latent heat of fusion of a pure A, k_e is the equilibrium partition coefficient, and m_L^e is the equilibrium liquidus slope. In this limit we also write $\Delta T = -|m_L^e| c_{\text{eq}}^L$. We thus have

$$
\Delta T\,(s_\alpha - s_L) = (RT/\Omega)\; c_{\text{eq}}^L\,(1 - k_e) \tag{13.9}
$$

Next, equating the equilibrium chemical potentials of solid and liquid using Eq. (13.7) ($\mu_\vartheta^{\text{eq}} = \partial f_\vartheta(c_{\text{eq}}^\vartheta)/\partial c$, $\vartheta = \alpha, l$), gives, to first order in concentration,

$$
(\mathcal{E}_\alpha - \mathcal{E}_L)\, c^L \approx -(RT/\Omega) c^L \ln k_e \tag{13.10}
$$

Substituting Eqs. (13.9) and (13.10) into Eq. (13.8) and re-scaling by RT/Ω, allows us to re-write Eq. (13.6) as

$$
c^L \ln\left[\frac{k(v_0)}{k_e}\right] + c^L\,(1 - k(v_0)) - c_{\text{eq}}^L\,(1 - k_e) = -\frac{v_0}{v_c}, \tag{13.11}
$$

where we defined a *velocity-dependent* segregation coefficient by

$$
k(v_0) = c^\alpha / c^L \tag{13.12}
$$

Re-writing $c^L_{\mathrm{eq}} = (T - T_A)/m^e_L$ in Eq. (13.11), and re-arranging, finally gives an expression for the interface temperature as

$$T = T_A + \frac{m^e_L}{1-k_e}\left\{\ln\left[\frac{k(v_0)}{k_e}\right] + [1 - k(v_0)]\right\} c^L + \frac{m^e_L\, v_0}{(1-k_e)\, v_c} \tag{13.13}$$

Equation (13.13) recovers the normal equation for the liquidus line of an ideal binary alloy when $v_0 \to 0$ and $k(v_0) \to k_e$. However, for non-zero velocities, this equation leads to a "kinetic phase diagram", described by a velocity dependent liquidus and solidus. For Eq. (13.13) to be meaningful we also require a model for $k(v_0)$. This has been addressed in numerous papers, and will also be addressed below in the context of a phase field model.

The continuous growth model of Aziz and Boettinger presented in Ref. [54] considered a slightly different driving force than Eq. (13.2), given by $\Delta G_{\mathrm{eff}} = \Delta G_{\mathrm{sol}} - \mathcal{D}\Delta G_{\mathrm{drag}}$, where $\mathcal{D}$ is a phenomenological parameter called the *solute drag coefficient.* The case $\mathcal{D} = 0$ corresponds to zero solute drag at the interface, while $\mathcal{D} = 1$ includes full solute drag, which is the case considered above. For the case of non-zero ($\mathcal{D} \neq 0$), it is straightforward to show that substituting Eq. (13.7) into $\Delta G_{\mathrm{eff}} = \Delta G_{\mathrm{sol}} - \mathcal{D}\Delta G_{\mathrm{drag}}$, and using Eq. (13.3) and Eq. (13.4), gives [54]

$$\begin{aligned} T = T_A + &\frac{m^e_L}{1-k_e}\left\{\Big[k(v_0) + [1 - k(v_0)]\,\mathcal{D}\Big]\ln\left[\frac{k(v_0)}{k_e}\right] + \Big[1 - k(v_0)\Big]\right\} c^L \\ &+ \frac{m^e_L\, v_0}{(1-k_e)\, v_c} \end{aligned} \tag{13.14}$$

It is noteworthy that the interface undercooling proposed by Eq. (13.13) was also obtained by Galenko and co-workers [95], but only in the limit when non-inertial dynamics of diffusing atoms are considered in the liquid. As a result, Eq. (13.13) or Eq. (13.14) are only expected to describe non-equilibrium effects at the lower range of interface velocities associated with the rapid solidification regime.

13.2 CONTINUOUS GROWTH MODEL LIMIT OF THE GRAND POTENTIAL PHASE FIELD MODEL

For this section, readers are referred to Appendix (B) for the general calculation of the asymptotic analysis of a binary alloy, which is applicable to Eqs. (6.6) and (6.9) in the limit of a binary alloy. This calculation examines the connection between asymptotically expanded solutions of the phase field equations within the interface (Eq. B.11) to the asymptotically expanded solutions far from the interface (Eq. B.10), in the limit $\epsilon = W_\phi/d_o \ll 1$, where d_o the capillary length of an alloy. In this section we analyze two specific equations obtained in Appendix (B) that relate corrections of the outer chemical potential, evaluated at (i.e projected into) the interface, to the curvature and normal velocity of the interface. The first equation is obtained at $\mathcal{O}(\epsilon)$ in the perturbation expansion and second is obtained at $\mathcal{O}(\epsilon^2)$.

The $\mathcal{O}(\epsilon)$ analysis of the phase field equation gives Eq. (B.43), namely

$$-(\bar{D}\bar{v}_0 + \bar{\kappa})\sigma_\phi - \Delta\omega(\mu_0^o(0^\pm)) = 0, \tag{13.15}$$

where $\bar{D} = D_{L\tau}/W_\phi^2$, σ_ϕ is given by Eq. (B.41) (with ϕ_o^{in} being the solution of Eq. B.24), $\bar{\kappa}$ is the dimensionless interface curvature, $\bar{v}_0$ is the $\mathcal{O}(1)$ correction of the dimensionless interface velocity, $\Delta\omega(\mu) \equiv \left[\omega^\alpha(\mu) - \omega^l(\mu)\right]/\alpha$ is the dimensionless grand potential density of the solid minus that of the liquid, and its argument $\mu_0^o(0^\pm)$ is the $\mathcal{O}(1)$ component of the outer chemical potential, evaluated at the solid (0^-)/liquid(0^+) side of the interface. *It is found that* $\mu_0^o(0^+) = \mu_0^o(0^-)$, and recalled that velocity is scaled by $v_s = D_L/d_o$, curvature by d_o, and $\alpha \propto RT/\Omega$ where the proportionality constant is not important here since it factors out of all equations.

The $\mathcal{O}(\epsilon^2)$ analysis of the phase field equation gives Eq. (B.69), where at this order (in ϵ) $\mu_1^o(0^+) \neq \mu_1^o(0^-)$. This equation takes on two forms on either side of the solid or liquid side of interface, namely,

$$-\bar{D}\sigma_\phi\bar{v}_1 - \frac{\Delta\tilde{c}(\mu_0^o(0^\pm))}{\alpha}\left(K + F^\pm\right)\bar{v}_0 + \frac{\Delta\tilde{c}(\mu_0^o(0^\pm))}{\alpha}\mu_1^o(0^\pm) = 0, \tag{13.16}$$

where $\bar{v}_1$ is the $\mathcal{O}(\epsilon)$ correction of the dimensionless interface velocity expansion, $\Delta\tilde{c}(\mu) \equiv \left[c^\alpha(\mu) - c^L(\mu)\right]$, $F^\pm$ and K are defined in Eq. (B.59) and Eq. (B.68), respectively, and $c^\alpha(c^L)$ are the equilibrium solid(liquid) concentrations, which here are functions of the $\mathcal{O}(1)$ chemical potential at the interface.[4] We will see below that considering the solid-side of Eq. (13.16) will correspond to modelling the CGM equation in Eq. (13.6) i.e., full drag, while considering the liquid side corresponds to considering the case of zero drag.

[4]To keep notation consistent with this section, we re-write the superscript in the appendix symbol $c^l(\mu)$ as $c^L(\mu)$ in this section.

Adding Eq. (13.15) to $\epsilon \times$ Eq. (13.16) gives

$$-\bar{D}\sigma_\phi \bar{v}_0 - \frac{\Delta\tilde{c}(\mu_0^o(0^\pm))}{\alpha}\left(K + F^\pm\right)[\epsilon\bar{v}_0] - \bar{D}\sigma_\phi\,[\epsilon\bar{v}_1]$$
$$-\frac{\Delta\omega(\mu_0^o(0^\pm))}{\alpha} + \frac{\Delta\tilde{c}(\mu_0^o(0^\pm))}{\alpha}\,[\epsilon\mu_1^o(0^\pm)] = 0, \qquad (13.17)$$

where for simplicity we re-define $\Delta\omega(\mu) \equiv \left[\omega^\alpha(\mu) - \omega^l(\mu)\right]$ for the remainder of this section. Here, we have dropped the curvature term since in rapid solidification $d_o\kappa \ll 1$, and most works consider a planar interface when considering the kinetics of rapid solidification. It is instructive to rescale Eq. (13.17) back to dimensional coordinates, done by noting that $\epsilon = W_\phi/d_o$ and $d_o = W_\phi/(\alpha\lambda)$, while the characteristic velocity is $v_s = d_o/D_L$. This gives $\bar{D}\sigma_\phi\bar{v}_0 = (\tau\sigma_\phi/\alpha W_\phi\lambda)\,v_0$, $\epsilon\bar{v}_0 = (W_\phi/D_L)\,v_0$ and $\epsilon\bar{v}_1 = (\tau\sigma_\phi/\alpha W_\phi\lambda)\,\epsilon\, v_1$. With these re-scalings, Eq. (13.17) is simplified to the form

$$-\frac{\tau\sigma_\phi}{W_\phi\lambda}\left\{1 - \frac{\lambda\Delta c(\mu_0^o(0^\pm))}{\sigma_\phi\bar{D}}\left(K + F^\pm\right)\right\} v_0$$
$$-\Delta\omega(\mu_0^o(0^\pm)) - \Delta c(\mu_0^o(0^\pm))\,\epsilon\mu_1^o(0^\pm) = 0, \qquad (13.18)$$

where $\Delta c(\mu) = c^L(\mu) - c^\alpha(\mu)$ as defined in Eq. (B.8). In arriving at Eq. (13.18), we dropped the ϵv_1 term in Eq. (13.17) as it is $< \mathcal{O}(\epsilon)$ and can thus be ignored up to order of ϵ used to construct Eq. (13.18) (see also discussion after Eq. B.71).

We simplify $\Delta\omega(\mu_0^o(0^\pm))$ in Eq. (13.18) by noting that in coexistence, $\mu_0^o(0^\pm)$ corresponds to two values of concentration given by $c_0^{\text{in}}(\xi \to -\infty) = c^\alpha(\mu_0^o(0^\pm))$ and $c_0^{\text{in}}(\xi \to \infty) = c^L(\mu_0^o(0^\pm))$, where $c_0^{\text{in}}(\xi)$ is the $\mathcal{O}(1)$ concentration in Eq. (B.30). This allows us to write

$$\Delta\omega(\mu_0^o(0^\pm)) = f_\alpha\left(c^\alpha\left(\mu_0^o(0^\pm)\right)\right)$$
$$- f_L\left(c^L\left(\mu_0^o(0^\pm)\right)\right) + \mu_0^o(0^\pm)\left[c^L\left(\mu_0^o(0^\pm)\right) - c^\alpha\left(\mu_0^o(0^\pm)\right)\right], \qquad (13.19)$$

where $f_L(c)$ and $f_\alpha(c)$ are the free energy density of the liquid and solid, respectively. Substituting Eq. (13.19) into Eq. (13.18) and re-arranging terms gives

$$\left[c^L\left(\mu_0^o(0^\pm)\right) - c^\alpha\left(\mu_0^o(0^\pm)\right)\right]\mu^o(0^\pm) = f_L\left(c^\alpha\left(\mu_0^o(0^\pm)\right)\right)$$
$$- f_\alpha\left(c^L\left(\mu_0^o(0^\pm)\right)\right) - \frac{\tau\sigma_\phi}{W_\phi\lambda}\left\{1 - \frac{\lambda\Delta c(\mu_0^o(0^\pm))}{\sigma_\phi\bar{D}}\left(K + F^\pm\right)\right\} v_0, \qquad (13.20)$$

where $\mu^o(0^{\pm}) = \mu_0^o(0^{\pm}) + \epsilon\mu_1^o(0^{\pm})$ is the total outer chemical potential on the solid $(-)$ or liquid $(+)$ sides of the interface, re-summed to order ϵ. (It is recalled that the lowest order chemical potential $\mu_0^o(0^{\pm})$ is equal on both sides of the interface, while F^+ and F^- are not). Equation (13.20) relates the outer chemical potential on the either side of an effective sharp interface (defined in the inner region of phase field equations) to the difference in free energy and difference in concentration on the respective sides of the interface (each related to the lowest order chemical potential), and to the interface velocity v_0.

13.2.1 Specializing Eq. (13.20) into the CGM Model of Eq. (13.6): Full Drag Case

To proceed, we consider the solid side of Eq. (13.20), and express the chemical potential $\mu^o(0^-)$ via the free energy density as

$$\mu^o(0^-) = \left.\frac{\partial f_\alpha(c)}{\partial c}\right|_{c=c^\alpha(\mu^o(0^-))} \approx \frac{\partial f_\alpha(c^\alpha(\mu_0^o(0^{\pm})))}{\partial c}, \tag{13.21}$$

where we approximate the concentration using the lowest order of the chemical potential. In what follows for the remainder of this section, we re-scale energy density according to $\bar{f}_\vartheta = f_\vartheta/(RT/\Omega)$, where $\vartheta = \alpha,\ L$, and Ω represents the molar volume[5] of the material. These considerations simplify the solid-side evaluation of Eq. (13.20) to

$$\bar{f}_\alpha\left(c^\alpha\right) - \bar{f}_L\left(c^L\right) + \left(c^L - c^\alpha\right)\frac{\partial \bar{f}_\alpha(c^\alpha)}{\partial c} = -\frac{v_0}{v_c^{\mathrm{PF}}}, \tag{13.22}$$

where we defined an inverse critical velocity $1/v_c^{\mathrm{PF}}$ by

$$\frac{1}{v_c^{\mathrm{PF}}} = \frac{\sigma_\phi\,\tau}{\hat{\lambda}W_\phi}\left\{1 - \frac{\hat{\lambda}(\Delta c)^2}{\sigma_\phi\,\bar{D}\,\bar{\chi}^L}\left(\bar{K} + \bar{F}^-\right)\right\}, \tag{13.23}$$

and where $\hat{\lambda} = RT/(\Omega\,H)$, H the inverse nucleation energy, $\bar{\chi}^L = (RT/\Omega)\,\partial c^L(\mu_0^o(0^{\pm}))/\partial\mu$, and the coefficients $\bar{K}, \bar{F}^-$ are *dimensionless versions* of the constants K, F^-, which are calculated in the next section. Equation (13.22) is the same form as Eq. (13.6). For the case of

[5]This re-scaling considers the energy density is in units of J/m^3 rather than J/mol as is commonly done in materials science. An equivalent expression for the energy scale is $k_B T\bar{\rho} = RT/\Omega$, where $\bar{\rho}$ is the number density of the material, which we assume here is the same in solid and liquid.

ideal binary alloys given by Eq. (13.7), Eq. (13.22) thus predicts a kinetic undercooling described by Eq. (13.13) (or alternatively Eq. 13.14, with $\mathcal{D} = 1$) with an effective $v_c \rightarrow v_c^{\rm PF}$ and some $k(v_0)$, the specific form of which will be determined in the next section.

13.2.2 Specializing Eq. (13.20) into the CGM Model with Zero Drag

We next consider the liquid side of Eq. (13.20) and express the liquid-side chemical potential $\mu^o(0^+)$ as

$$\mu^o(0^+) = \left.\frac{\partial f_L(c)}{\partial c}\right|_{c=c^L(\mu^o(0^+))} \approx \frac{\partial f_L(c^L(\mu_0^o(0^\pm)))}{\partial c} \tag{13.24}$$

Using Eq. (13.24), the liquid-side evaluation of Eq. (13.20) now becomes

$$\bar{f}_\alpha\left(c^\alpha\right) - \bar{f}_L\left(c^L\right) + \left(c^L - c^\alpha\right)\frac{\partial \bar{f}_L(c^L)}{\partial c} = -\frac{v_0}{v_c^{\rm PF}}, \tag{13.25}$$

where now the effective inverse critical velocity $1/v_c^{\rm PF}$ is defined by

$$\frac{1}{v_c^{\rm PF}} = \frac{\sigma_\phi \tau}{\hat{\lambda} W_\phi}\left\{1 - \frac{\hat{\lambda}(\Delta c)^2}{\sigma_\phi \bar{D}\,\bar{\chi}^L}\left(\bar{K} + \bar{F}^+\right)\right\}, \tag{13.26}$$

where $\bar{F}^+$ is a *dimensionless version* of the constant, F^+, which will be calculated in the next section. All other variables in Eq. (13.26) are as described in the previous section.

It is straightforward to show that the form of Eq. (13.25) corresponds identically to the case of $\Delta G_{\rm eff} = \Delta G_{\rm sol}$ (i.e. $\Delta G_{\rm drag} = 0$) [54]. As a result, tuning the effective critical velocity of the phase field model to $v_c^{\rm PF}$ given by Eq. (13.26) corresponds to the simulating Eq. (13.14) for the case of zero solute drag, i.e., $\mathcal{D} = 0$.

13.2.3 Relating $1/v_c^{\rm PF}$ to Interface Kinetic Coefficient β for the Case of Ideal Binary Alloys

The CGM model of Eq. (13.13) is a sharp interface model for an ideal binary alloy. Inspection of the last term yields

$$\frac{1}{\mu} = \frac{m_L^e}{(1-k_e)v_c}, \tag{13.27}$$

where μ is the effective interface mobility, which for a binary alloy is related to an interface kinetic coefficient β according to $1/\mu = \beta \Delta T_o$, where $\Delta T_o = (1 - k_e)c_o^L$ is the freezing range of the alloy [10] and c_o^L is a reference equilibrium concentration. Equation (13.27) can thus be written as

$$\frac{1}{v_c} = (1 - k_e)^2 \, c_o^L \, \beta \tag{13.28}$$

Combining Eq. (13.28) with either Eq. (13.23) or Eq. (13.26) gives (for an ideal binary alloy),

$$\beta^{\pm} = \frac{\sigma_\phi \, \tau}{\lambda \, J W_\phi} \left\{ 1 - \frac{\lambda}{\bar{D}} \frac{J \left(\bar{K} + \bar{F}^{\pm} \right)}{\sigma_\phi} \right\}, \tag{13.29}$$

where $J = 16/15$ and where λ is given by

$$\lambda = \frac{15RT(1 - k_e)^2 c_o^L}{16\Omega H} \tag{13.30}$$

In arriving at Eq. (13.29), we made the approximation $\Delta c \approx \Delta c_F$ and $\bar{\chi}^L \approx \bar{\chi}^{L(eq)}$ for simplicity.[6]

The kinetic coefficient in Equation (13.29) takes on different values of the solid or liquid side of the interface depending on the values of $\bar{F}^+$ or $\bar{F}^-$. The choice of β, in turn, controls the kinetic time scale of the phase field equations. Thus, tuning the parameters of the phase field equations to cater to the choice of β^+ versus β^- corresponds to effectively mapping the phase field model onto the two CGM cases considered above, one with full drag (β^-) and the other with no drag (β^+). It is plausible that considering a mixture of the two β values can lead to intermediate forms of drag described by $\mathcal{D}$, but that is not considered here.

13.3 NON-EQUILIBRIUM PARTITION COEFFICIENT $k(V_0)$ AND CHOICE OF ANTI-TRAPPING

This section uses results from the asymptotic analysis in Appendix (B) to derive the form of the non-equilibrium partition coefficient $k(v_0)$ needed to compute solute partitioning and interface undercooling in the CGM

[6] It is recalled that Δc_F is the equilibrium concentration difference at the interface and $\bar{\chi}^{L(eq)}$ the susceptibility evaluated at the equilibrium chemical potential. A more accurate expression for β contains a correction c_L/c_o^L multiplying the $(\bar{K} + \bar{F})$ term, which would require that the kinetic time scale τ for the order parameter equation have a correction similar to that discussed in Refs. [10, 38].

model (i.e., Eq. 13.14). Along the way, we will also compute the parameters required $\{\bar{F}^+, \bar{F}^-, \bar{K}\}$, which also appear in the kinetic coefficient. We will see that this requires the introduction of a new form for the anti-trapping current from that used in phase field models that emulate the classic sharp SIM limit discussed earlier.

13.3.1 Chemical Potential Jump at the Interface

The derivation of $k(v_0)$ begins by considering Eq. (B.60) in the asymptotic analysis, which relates the [first order] chemical potential jump at the interface to the dimensionless velocity according to $\mu_1^o(0^+) - \mu_1^o(0^-) = (F^+ - F^-)\bar{v}_0$, where F^+ and F^- are given in Eq. (B.59). Multiplying both side of this equation by ϵ, recalling that $\epsilon = W_\phi v_s/D_L$ and $\bar{v}_0 = v_0/v_s$, and noting that $\mu_0^o(0^\pm)$ is the same on both sides of the interface, gives

$$\mu^o(0^+) - \mu^o(0^-) = \frac{W_\phi\, \Delta c \Delta \bar{F}}{D_L\, \chi^L}\, v_0, \tag{13.31}$$

where $\Delta\bar{F} = \bar{F}^+ - \bar{F}^-$ with $\bar{F}^+$, $\bar{F}^-$ being the constants referenced previously, given explicitly by

$$\begin{aligned}\bar{F}^+ &= \int_0^\infty \left\{1 - \frac{[c_0^{\text{in}}(x) - c_s]}{\Delta c\, \tilde{q}(\phi_0^{\text{in}})\, \bar{\chi}(\phi_0^{\text{in}})}\right\} dx + \int_0^\infty \frac{a_t(\phi_0^{\text{in}})}{\tilde{q}(\phi_0^{\text{in}})\, \bar{\chi}(\phi_0^{\text{in}})} \left(\frac{d\phi_0^{\text{in}}}{dx}\right) dx \\ \bar{F}^- &= \int_{-\infty}^0 \frac{[c_0^{\text{in}}(x) - c_s]}{\Delta c\, \tilde{q}(\phi_0^{\text{in}})\bar{\chi}(\phi_0^{\text{in}})} dx - \int_0^\infty \frac{a_t(\phi_0^{\text{in}})}{\tilde{q}(\phi_0^{\text{in}})\, \bar{\chi}(\phi_0^{\text{in}})} \left(\frac{d\phi_0^{\text{in}}}{dx}\right) dx,\end{aligned} \tag{13.32}$$

where the expressions appearing in Eq. (13.32) are defined as follows: *(1)* the lowest order concentration $c_0^{\text{in}}(x)$ (comes from Eq. B.30) is

$$c_0^{\text{in}}(x) = c_L \left\{1 - (1-k)\, h(\phi_0^{\text{in}}(x))\right\}, \tag{13.33}$$

where $k = c_s/c_L$ is the partition coefficient, with $c_s = c^\alpha(\mu_0^o(0^\pm))$, $c_L = c^L(\mu_0^o(0^\pm))$ defined as in Eq. (B.39)[7]; *(2)* the dimensionless susceptibility function $\bar{\chi}(\phi_0^{\text{in}})$ is given by

$$\bar{\chi}(\phi_0^{\text{in}}) = \left\{1 - \left(1 - k^{\text{eff}}\right) h(\phi_0^{\text{in}}(x))\right\}, \tag{13.34}$$

[7]It is recalled from the discussion above that the superscript in the appendix symbol $c^l(\mu)$ is expressed in this section as $c^L(\mu)$.

where $k^{\text{eff}} = \chi^\alpha/\chi^L$, with $\chi^L = \partial c^L(\mu_0^o(0^\pm))/\partial\mu$ and $\chi^\alpha = \partial c^\alpha(\mu_0^o(0^\pm))/\partial\mu$ (it is noted that for an ideal alloy described by Eqs. (13.7) $k^{\text{eff}} = k$); *(3)* The interpolation function of the diffusivity $\tilde{q}(\phi_0^{\text{in}})$ is chosen to have the specific form

$$\tilde{q}(\phi_0^{\text{in}}) = \frac{\bar{q}(\phi_0)}{\{1-(1-k^{\text{eff}})\,h(\phi_0^{\text{in}}(x))\}}, \tag{13.35}$$

whose denominator cancels the $\bar{\chi}(\phi_0^{\text{in}})$ in the denominators of Eq. (13.32).

To proceed further, we choose the interpolation functions to be the commonly used forms $\bar{q}(\phi) = 1-\phi$ and $h(\phi) = \phi$, which it is straightforward to show simplifies $\bar{F}^+$ and $\bar{F}^-$ to

$$\begin{aligned} \bar{F}^+ &= \int_0^\infty \frac{a_t(\phi_0^{\text{in}})\,\partial_x\phi_0^{\text{in}}}{1-\phi_0^{\text{in}}}\,dx \\ \bar{F}^- &= \int_0^\infty \left\{1 - \frac{a_t(\phi_0^{\text{in}})\,\partial_x\phi_0^{\text{in}}}{1-\phi_0^{\text{in}}}\right\}dx \end{aligned} \tag{13.36}$$

It is also noted that with these choices of interpolation functions $\bar{K}$ in Eq. (13.23) and Eq. (13.26) becomes

$$\bar{K} = \int_{-\infty}^\infty \frac{\partial\phi_0^{\text{in}}}{\partial\xi}P'(\phi_0^{\text{in}})\left\{\int_0^\xi \frac{\left[(\phi_0^{\text{in}}-1)+a_t(\phi_0^{\text{in}})\,\partial_x\phi_0^{\text{in}}\right]}{1-\phi_0^{\text{in}}}dx\right\}d\xi \tag{13.37}$$

Note: *For the remainder of this chapter the "in" superscript for the lowest order order solutions of the fields will be dropped to simplify notation.* Choosing the double well potential function to be the often used form $g(\phi_0) = \phi^2(1-\phi_0)^2$ gives $d\phi_0/dx = -\sqrt{2}\phi_0\,(1-\phi_0)$, which further simplifies Eqs. (13.36) and (13.37) to

$$\begin{aligned} \bar{F}^+ &= -\sqrt{2}\int_0^\infty a_t(\phi_0)\phi_0(x)dx \\ \bar{F}^- &= \int_0^\infty \left\{1+\sqrt{2}\,a_t(\phi_0)\phi_0\right\}dx \\ \bar{K} &= -\int_{-\infty}^\infty \frac{\partial\phi_0^{\text{in}}}{\partial\xi}P'(\phi_0)\left\{\int_0^\xi \left[1+\sqrt{2}\,a_t(\phi_0)\phi_o\right]dx\right\}d\xi \end{aligned} \tag{13.38}$$

The interpolation function $P(\phi)$ used here will be such that $P'(\phi_0) = 30\,\phi_0^2\,(1-\phi_0)^2$.

The expression $\Delta\bar{F}$ emerges as one of the so-called "corrections" that spoils the exact correspondence between the diffuse interface limit of the phase field equations and the classic SIM model. Contact with the limit of the classic SIM model requires that $\Delta\bar{F} = 0$, which it is straightforward to show is true for the choice $a_t = -1/\sqrt{2}$ [10, 38]. In contrast, contact with the CGM SIM limit will require that $\Delta\bar{F} \neq 0$. It is also noted here that two additional corrections also emerge, given by

$$\Delta H = c_L(1-k)\left(\int_0^{\infty} \phi_0(x)\, dx - \int_{-\infty}^{0} [1-\phi_0(x)]\, dx\right) \tag{13.39}$$

and $\Delta J = \chi^L \Delta H$. For the above choices of interpolation functions, it is straightforward to show that $\Delta H = \Delta J = 0$, regardless of the choice of $a_t(\phi_0)$.

To proceed further, it will be necessary to choose a form for the anti-trapping flux $a_t(\phi_0^{\text{in}})$ in Eq. (13.38) such as to achieve a specific *non-zero* $\Delta\bar{F}$. Following this, we will use the free energy of the specific alloy under study in Eq. (13.31) in order to evaluate $k(v_0)$ as a function of $\Delta\bar{F}$. The specific choice of $a_t(\phi_0^{\text{in}})$ will also determine $\bar{F}^{\pm}$ and $\bar{K}$ entering $v_c^{;\text{PF}}$ in Eq. (13.23) or Eq. (13.26) (or equivalently β^+/β^- in Eq. 13.29).

13.3.2 Evaluation of $\Delta\bar{F}$ and an Implicit Equation for $k(v_0)$ from Eq. (13.31)

To achieve a controlled amount of solute trapping, the anti-trapping flux $a_t(\phi)$ is modified to the form

$$a_t(\phi) = -\frac{1}{\sqrt{2}}\left\{1 - A\phi\,(1-\phi)\right\}, \tag{13.40}$$

where A is a constant[8] that we refer to as the *solute trapping parameter*[9]. Substituting Eq. (13.40) into the each of the equations in Eq. (13.38) and carrying out the integrals with the above-specified interpolation

[8]It is recalled that the sign of $a_t(\phi)$ was changed in the appendix for convenience. We must thus rescale the anti-trapping in Eq. (13.40) according to $a_t(\phi) \to -a_t(\phi)$ when inserted into any of the phase field models discussed earlier.

[9]This form of anti-trapping is just one of numerous classes of functions that can be used.

functions gives

$$\Delta\bar{F} = -\frac{\sqrt{2}\,A}{4} \tag{13.41}$$

$$\bar{F}^- = \frac{\sqrt{2}\ln 2}{2} + \frac{3\sqrt{2}A}{16} \tag{13.42}$$

$$\bar{K} = 0.0638 - 0.01263\,A \tag{13.43}$$

Readers using the alloy model with the order parameter defined in the range $-1 < \phi < 1$ can calculate the above integrals with the transformation $\phi_o \to (\phi_o + 1)/2$, which results in $P'(\phi_0) = (15/16)\left(1 - \phi_0^2\right)^2$, while $g(\phi_0) = -\phi_0^2/2 + \phi_0^4/4$ and $d\phi_0/dx = -\left(1 - \phi_0^2\right)/\sqrt{2}$. This then gives

$$\Delta\bar{F} = -\sqrt{2}A \tag{13.44}$$

$$\bar{F}^- = \frac{\sqrt{2}\ln 2}{2} + \frac{3\sqrt{2}A}{4} \tag{13.45}$$

$$\bar{K} = 0.0638 - 0.0505\,A \tag{13.46}$$

To determine $k(v_0)$ as a function of A, we use Eq. (13.7) for Henrian alloys in Eq. (13.31), and use Eq. (13.41) for $\Delta\bar{F}$. This gives

$$\ln\left[\frac{k(v_0)}{k_e}\right] = \left(\frac{A}{2\sqrt{2}}\right)\{1 - k(v_0)\}\,\frac{v_0}{v_D^{\mathrm{PF}}}, \tag{13.47}$$

where

$$v_D^{\mathrm{PF}} = \frac{D_L}{W_\phi} \tag{13.48}$$

13.3.3 Computing $k(v_0)$ for an Ideal Binary Alloy

To specialize Eq. (13.47) to an ideal binary alloy, it is useful to eliminate W_ϕ in v_D^{PF} in terms of d_o using the asymptotic result $W_\phi = d_o\bar{\lambda}/a_1$, where

$$\bar{\lambda} = \frac{RT(1 - k_e)^2\, c_L^{\mathrm{eq}}}{\Omega J\, H} \tag{13.49}$$

is the coupling constant used in the dimensionless phase field equations, where $J = 16/15$ for the case where $0 < \phi < 1$ (or $-1 < \phi < 1$). This gives

$$v_D^{\mathrm{PF}} = \left(\frac{4a_1}{\sqrt{2}\,\bar{\lambda}A}\right)\frac{D_L}{d_o} \tag{13.50}$$

Since in phase field simulations W_ϕ takes on mesoscale values for computational efficiency, we use the symbol δ to denote the *physical* interface with. With a value of $\delta \sim 10^{-9}$ m, this gives $v_D \sim D_L/\delta \sim 1$ m/s in metals. In dilute alloys, it is found that $\delta = f d_o$, where typically $0.1 < f < 1$. These considerations transform Eq. (13.47) to

$$\ln\left[\frac{k(v_0)}{k_e}\right] = \left(\frac{\sqrt{2}|A|\bar{\lambda}}{4a_1 f}\right)\{1 - k(v_0)\}\frac{v_0}{v_D^{PF}} \tag{13.51}$$

The remarkable feature of Eq. (13.51) is that, once a specific ratio $\bar{\lambda} = W_\phi/d_o$ is chosen to set the scale of the phase field interface (W_ϕ), we can then use the degree of freedom of the solute trapping parameter (A) to achieve a desired $k(v_0)$ corresponding to a specific v_D.

To complete the quantitative mapping of the phase field model onto the continuous growth model in Eq. (13.13), we need to choose the time scale τ to correspond to a desired atomic attachment speed v_c. This can be done by solving Eq. (13.23) for τ once $v_c^{\rm PF}$ is specified (e.g. following Boettinger et al. we use the speed of sound). To do so, it is useful to express Eq. (13.23) in terms of the $\bar{\lambda}$, which becomes

$$v_c^{\rm PF} = \frac{1}{a_1^2(1-k_e)^2 c_L^{\rm eq}}\frac{d_o\bar{\lambda}^2}{\tau\,\beta}, \tag{13.52}$$

where the kinetic coefficient in Eq. (13.52), also expressed in terms of $\bar{\lambda}$, is given (to lowest order in velocity) by

$$\beta = 1 - \frac{a_2}{a_1^2}\frac{d_o^2}{D_L}\frac{\bar{\lambda}^3}{\tau}, \tag{13.53}$$

where

$$\begin{aligned} a_1 &= \frac{\sigma_\phi}{J} \\ a_2 &= J\left(\frac{\bar{K} + \bar{F}^-}{\sigma_\phi}\right) \\ \sigma_\phi &= \int_{-\infty}^{\infty}(\partial_x\phi_o)^2\,dx \end{aligned} \tag{13.54}$$

The constant $\sigma_\phi = \sqrt{2}/6$ for $0 < \phi < 1$ and $\sigma_\phi = 2\sqrt{2}/3$ for the case $-1 < \phi < 1$. Figure (13.1) shows the computed $k(v_0)$ using $A = 0.2$ and $\bar{\lambda} = 2.5$ for an Al-3wt%Cu alloy, for which $k_e = 0.15$, $D_L = 3 \times 10^{-9}$ m^2/s and $d_o = 5.4 \times 10^{-9}$ m. For this alloy, the characteristic numerical interface length and kinetic time scales are set to $W_\phi = 1.6 \times 10^{-8}$ m and $\tau = 1.7 \times 10^{-7}$ s, respectively[10].

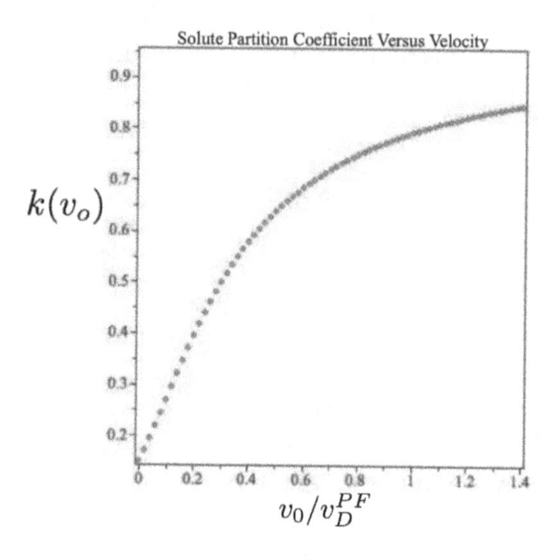

Figure 13.1 $k(v_0)$ versus v_0 using $|A| = 0.2$ in the anti-trapping function.

The above considerations assume isothermal conditions. For extensions of the above theory that consider non-isothermal conditions (e.g. directional solidification), the reader is referred to Refs. [96, 97], where the above formalism is extended. For the case of directional solidification it is noted that a U-dependent correction to the kinetic coefficient can also be applied to improve convergence to the sharp interface limit described by CGM theory; this is exactly analogous to the U-dependent correction discussed in Ref. [10] for convergence to the classical SIM limit for non-isothermal situations.

[10]It is noted that the above analysis tacitly assumes that the expansion parameter $W_\phi/d_o \sim 1$ in order for the results of the asymptotic analysis to remain valid. This does not pose too severe a restriction since the time scales for rapid solidification are on the scale of μs-ms, which makes simulations with "diffuse" interfaces even of order nm quite feasible.

CHAPTER 14

Application: Phase Field Simulations of Rapid Solidification of a Binary Alloy

THIS CHAPTER DEMONSTRATES THE RESULTS of CGM analysis presented in Chapter 13 for a system comprising a single crystal in an ideal binary alloy, which was first published in Ref. [96]. This system was modelled using the grand potential field model in Section 9.1.3, which in this case reduces the same form of the phase field equations derived in Sections 9.1.4 or 9.2 with a single order parameter. The model was parameterized to simulate a specific form of $k(v_o)$ and kinetic undercooling (corresponding to a kinetic interface concentration), as described by the CGM. Both limits of CGM (full drag and zero drag) are considered. For comparison, we also show the case of equilibrium partitioning $k(v_o) = k_e$ with kinetic coefficient β set to either zero or nonzero.[1]

To determine an appropriate amount of solute trapping in the phase field model, the characteristic solute trapping velocity of the model, v_D, was adjusted to match the $k(v)$ according to Eq. (13.51) for an experimentally fitted partition function $k(v)$ as closely as possible at low interface velocities. For details on all other parameters, the reader

[1]To keep notation as simple as possible, we will in this chapter use the notation v instead of v_o used in the previous chapter and so $k(v_o)$ will be referenced simply as $k(v)$.

DOI: 10.1201/9781003204312-14

is referred to Ref. [96]. It is recalled that a velocity dependent solute partitioning signifies that the interface is out of equilibrium. Under the conditions of the non-equilibrium CGM theory, the liquid-side interface concentration of the solidification front then obeys

$$\frac{c_L}{c_l^o} = \frac{1}{f\left(k(v)\right)}\left(1 + \frac{T_l - T}{|m_l^e| c_l^o} - \left(1 - k_e\right) d_o \kappa - \left(1 - k_e\right) \beta v\right), \tag{14.1}$$

where c_l^o is the average solute concentration in the alloy, T_l is the liquidus temperature, d_o is the solutal capillary length, κ is the local interface curvature, β is the kinetic coefficient, and $f\left(k(V)\right)$ is the velocity-dependent correction to the liquidus slope, given by

$$f\left(k(v)\right) = \frac{1}{1 - k_e}\left(\left[k(v) + \mathcal{D}(1 - k(v))\right] \log\left(\frac{k(v)}{k_e}\right) + 1 - k(v)\right), \tag{14.2}$$

where $\mathcal{D}$ is a parameter that can be tuned to represent complete solute drag ($\mathcal{D} = 1$) or no solute drag ($\mathcal{D} = 0$) [54].

Figure 14.1 shows the convergence of the partition coefficient $k(v)$ (left graph) and liquid-side concentration c_L (right graph) for the cases above. The data were obtained using two small computational interface widths W to demonstrate the mathematical convergence of the phase field equations to the aforementioned sharp interface theory.[2] The grey data corresponds to smaller interface width $W = 0.2$ nm, and black to $W = 0.5$ nm, and the dimensionless undercooling for these runs was set to $\Delta = 0.75$. As expected, the smaller interface width data (grey) converge to the corresponding CGM theory up to higher interface velocities than the larger interface width data (black). In all cases shown in Figure 14.1, the interface velocity decreases monotonically over time, and both the partition coefficient and the liquid-side concentration converge to the corresponding sharp interface model (solid and broken black lines) at the measured instantaneous velocities.

Figure 14.2 shows the same convergence as in Figure 14.1, except for material properties chosen to be for a Si-As alloy, and the phase field model's convergence onto the SIM limit of CGM theory is demonstrated using more diffuse interface widths than in the case of Figure 14.1, which is useful for increasing computational efficiency in 2D and 3D simulations. Scatter points in grey correspond to $W = 15$ nm, and in black to

[2]The reader is referred to the online version of Ref. [96] to discriminate the colours in the data points.

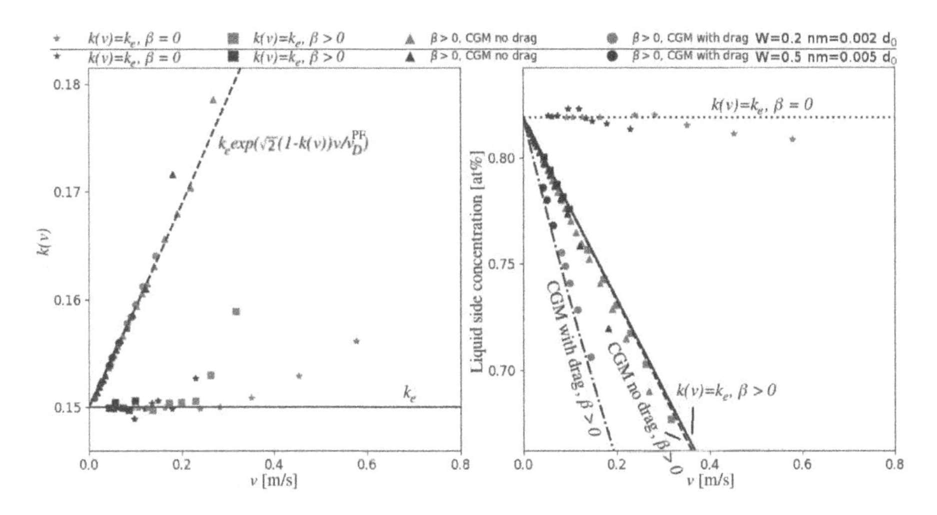

Figure 14.1 Convergence of different phase field simulations (grey and black scatter points) to corresponding sharp interface models (solid and broken black lines) for an Al-Cu alloy. The left graph shows convergence of the partition coefficient to k_e and $k(v)$ computed from Eq. (13.51). The right graph shows the convergence of the liquid-side concentration c_L to Eq. (14.1) for the different non-equilibrium cases indicated in the text. Each scatter point corresponds to an instantaneous velocity and interfacial concentration measurement taken from the transient evolving of the concentration profile, using dimensionless undercooling $\Delta = 0.75$ [96].

$W = 20$ nm; dimensionless undercooling is $\Delta = 0.55$. The phase field models converge well to the corresponding CGM (and classical) sharp interface models at low velocities. However, there is a larger relative scatter since the concentration projection error remains roughly the same as for the Al-Cu data in Figure 14.1, but velocities are smaller, corresponding to a partition coefficient $k(v)$ and liquid-side concentration $c_L(v)$ that are closer to the equilibrium values at $v = 0$. It is noted that the thin interface convergence to the sharp interface CGM here occurs over a lower range of speeds since the asymptotic analysis is most accurate at the lower driving forces. This range of velocities is relevant for industrial rapid solidification conditions such as laser welding.

Dynamic transmission electron microscopy (DTEM) experiments of thin film solidification were recently compared to multi-order parameter phase field simulations [97] in the limit of rapid solidification using

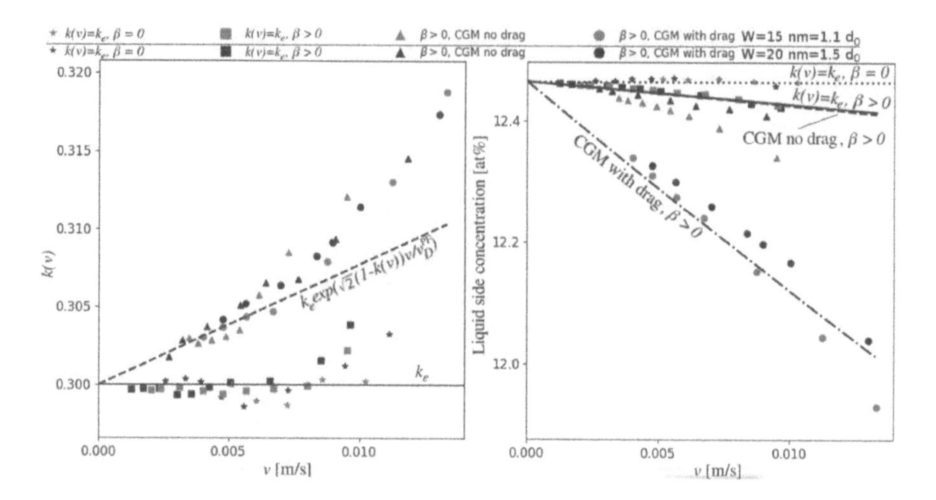

Figure 14.2 Convergence of different phase field simulations (grey and black scatter points) to corresponding sharp interface models (red and blue scatter points) for Si-Al alloy for two diffuse interface widths. The left graph shows convergence of the partition coefficient to k_e and $k(v)$ from Eq. (13.51). The right graph shows the convergence of the liquid-side concentration c_L to Eq. (14.1) for the different non-equilibrium cases indicated in the text. Dimensionless undercooling is set to $\Delta = 0.55$ [96].

the results of the previous section. The temperature distribution in the phase field model was described with a simple analytical model fitted to solid–liquid interface data from DTEM experiments. This temperature distribution was then used to drive the rapid solidification in phase field simulations. The phase field simulations and corresponding DTEM experiments are compared in Figure 14.3 for an Al-Cu alloy. The simulations and experiments are in fair agreement in terms of grain length scales and grain-boundary morphology (Figure 14.3a–b), and in terms of microsegregation magnitude and patterning (Figure 14.3c). However, the phase field simulations produce noticeably longer initial cells, as shown within the dashed vertical lines in Figure 14.3b. This discrepancy is likely due to the neglect of latent heat effects in the phase field simulations. Future studies combining phase field simulations and DTEM experiments are expected to form an important framework to test such assumptions related to heat-transfer conditions, as well as for parameterizing hard-to-measure kinetic parameters in phase field models, and, in general, to better understand the physics of rapid solidification.

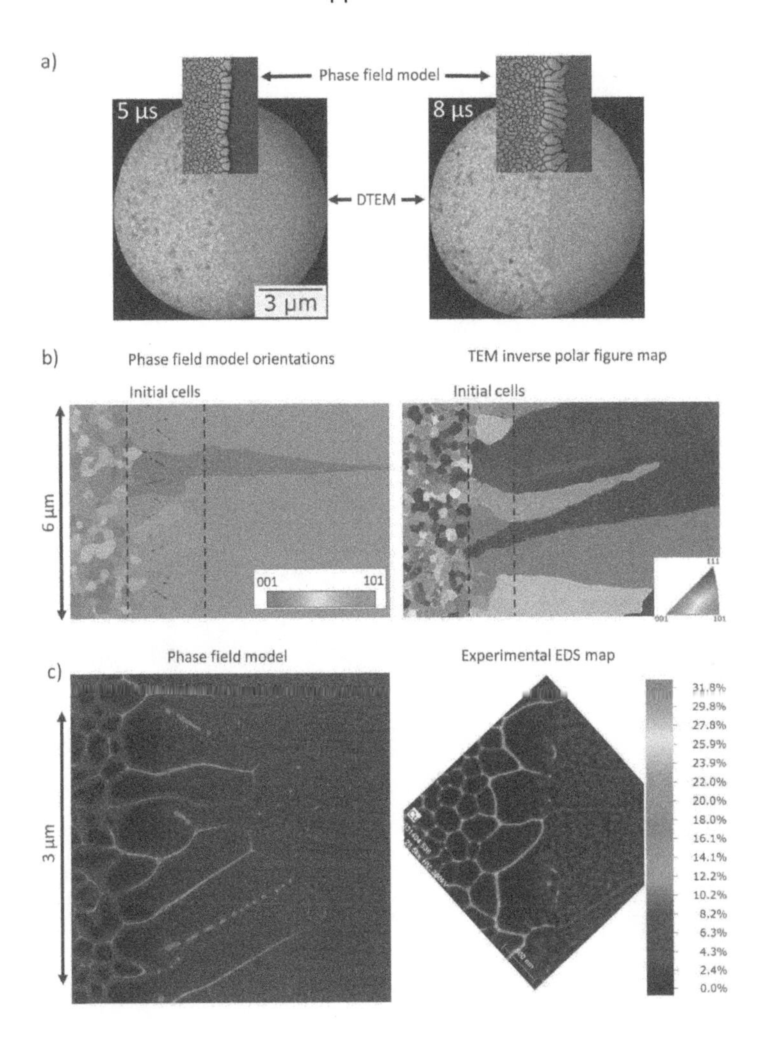

Figure 14.3 Rapid resolidification of a dilute Al-Cu in thin-film comparing phase field simulations and experiments. (a) phase field model and time-resolved dynamic transmission electron microscope (DTEM) microstructures at two similar times. Solid is to the left and liquid to the right; (b) orientation distribution, and (c) copper concentration obtained from energy-dispersive x-ray spectroscopy (EDXS), where lighter grey represents high segregation pockets of Cu and dark grey represents low concentration [97]. (Orientation and segregation maps are shown in the colour eBook.)

V

APPENDIX A

Incorporating Temperature in the Grand Potential Phase Field Model

TO INCLUDE THERMAL TRANSPORT into the grand potential phase field model, we re-write the grand potential functional to explicitly include temperature according to

$$\Omega[\boldsymbol{\phi}, \boldsymbol{\mu}, T] = \int_V \left\{ \omega_{\text{int}}\left(\boldsymbol{\phi}, \boldsymbol{\nabla}\boldsymbol{\phi}\right) + \sum_{\alpha}^{N} g_\alpha(\boldsymbol{\phi})\omega^\alpha(\boldsymbol{\mu}, T) + \left[1 - \sum_{\alpha}^{N} g_\alpha(\boldsymbol{\phi})\right] \omega^\ell(\boldsymbol{\mu}, T) \right\}, \tag{A.1}$$

where we have neglected the temperature (T) dependence on the interaction part of the functional. Strictly speaking, the surface term, energy barrier height, and grain interaction constants in ω_{int} should include some temperature dependence. However, these constants will be assumed for simplicity to be independent of temperature.

Temperature evolution is governed by the conservation of internal energy e, given by

$$\frac{\partial e}{\partial t} = \boldsymbol{\nabla}\left(k(\boldsymbol{\phi}, T)\boldsymbol{\nabla}T\right), \tag{A.2}$$

DOI: 10.1201/9781003204312-A

where $k(\boldsymbol{\phi}, T)$ is the thermal conductivity. Internal energy can be eliminated in favour of temperature through the entropy density, which is given by the variational of the grand potential with temperature by

$$s(\boldsymbol{\phi}, \boldsymbol{\mu}, T) = -\frac{\delta\Omega[\boldsymbol{\phi}, \boldsymbol{\mu}, T]}{\delta T}. \tag{A.3}$$

Equation (A.3) requires that we explicitly include temperature variation in the grand potential functional. This gives

$$\begin{aligned}
\delta\Omega = \int_V \Bigg(& \Bigg\{ -\sigma_\alpha \boldsymbol{\nabla}^2 \phi_\alpha + H_\alpha f'_{\mathrm{DW}}(\phi_\alpha) + \sum_{\beta \neq \alpha} \Psi'(\phi_\alpha, \phi_\beta) \\
& + g'_\alpha(\boldsymbol{\phi}) \left[\omega^\alpha(\boldsymbol{\mu}, T) - \omega^\ell(\boldsymbol{\mu}, T) \right] \Bigg\} \delta\phi_\alpha \\
& + \sum_i \left\{ \sum_\alpha^N g_\alpha(\boldsymbol{\phi}) \frac{\partial \omega^\alpha(\boldsymbol{\mu}, T)}{\partial \mu_i} + \left[1 - \sum_\alpha^N g_\alpha(\boldsymbol{\phi}) \right] \frac{\partial \omega^\ell(\boldsymbol{\mu}, T)}{\partial \mu_i} \right\} \delta\mu_i \\
& + \left\{ \sum_\alpha^N g_\alpha(\boldsymbol{\phi}) \frac{\partial \omega^\alpha(\boldsymbol{\mu}, T)}{\partial T} + \left[1 - \sum_\alpha^N g_\alpha(\boldsymbol{\phi}) \right] \frac{\partial \omega^\ell(\boldsymbol{\mu}, T)}{\partial T} \right\} \delta T \Bigg) \, d^3\boldsymbol{r}
\end{aligned} \tag{A.4}$$

From Eq. (A.4) we write the entropy density as

$$s(\boldsymbol{\phi}, \boldsymbol{\mu}, T) = -\left\{ \sum_\alpha^N g_\alpha(\boldsymbol{\phi}) \frac{\partial \omega^\alpha(\boldsymbol{\mu}, T)}{\partial T} + \left[1 - \sum_\alpha^N g_\alpha(\boldsymbol{\phi}) \right] \frac{\partial \omega^\ell(\boldsymbol{\mu}, T)}{\partial T} \right\}, \tag{A.5}$$

which can be cast into the form

$$s(\boldsymbol{\phi}, \boldsymbol{\mu}, T) = \sum_\alpha^N g_\alpha(\boldsymbol{\phi}) s^\alpha(\boldsymbol{\mu}, T) + \left[1 - \sum_\alpha^N g_\alpha(\boldsymbol{\phi}) \right] s^\ell(\boldsymbol{\mu}, T), \tag{A.6}$$

where we have defined $s^\vartheta = -\partial\omega^\vartheta / \partial T$, the entropy density for a particular phase. Analogously to the auxiliary concentration concentration fields, s^ϑ defines an *auxiliary* entropy density of phase ϑ. Using the thermodynamic relation $de = Tds$ (at constant pressure) gives

$$\frac{\partial e}{\partial t} = T \frac{\partial s}{\partial t} = T \left(\sum_\alpha^N \frac{\partial s}{\partial \phi_\alpha} \frac{\partial \phi_\alpha}{\partial t} + \sum_i \frac{\partial s}{\partial \mu_i} \frac{\partial \mu_i}{\partial t} + \frac{\partial s}{\partial T} \frac{\partial T}{\partial t} \right) \tag{A.7}$$

The terms of Eq. (A.7) can be derived from Eq. (A.6). We identify the third term as the specific heat capacity $c_p(\boldsymbol{\phi}, \boldsymbol{\mu}, T)$, given by

$$\begin{aligned} c_p(\boldsymbol{\phi}, \boldsymbol{\mu}, T) &= T\frac{\partial s}{\partial T} \\ &= \sum_{\alpha}^{N} g_\alpha(\boldsymbol{\phi}) c_p^\alpha(\boldsymbol{\mu}, T) + \left[1 - \sum_{\alpha}^{N} g_\alpha(\boldsymbol{\phi})\right] c_p^\ell(\boldsymbol{\mu}, T), \end{aligned} \tag{A.8}$$

where $c_p^\vartheta(\boldsymbol{\mu}, T) = T\,\partial s^\vartheta(\boldsymbol{\mu}, T)/\partial T$ defines the specific heat of phase ϑ. The middle term in Eq. (A.7) is given explicitly by

$$\frac{\partial s(\boldsymbol{\phi}, \boldsymbol{\mu}, T)}{\partial \mu_i} = \sum_{\alpha}^{N} g_\alpha(\boldsymbol{\phi}) \frac{\partial s^\alpha(\boldsymbol{\mu}, T)}{\partial \mu_i} + \left[1 - \sum_{\alpha}^{N} g_\alpha(\boldsymbol{\phi})\right] \frac{\partial s^\ell(\boldsymbol{\mu}, T)}{\partial \mu_i} \tag{A.9}$$

The first term in Eq. (A.7) becomes

$$\frac{\partial s(\boldsymbol{\phi}, \boldsymbol{\mu}, T)}{\partial \phi_\alpha} = \sum_{\alpha}^{N} g'_\alpha(\boldsymbol{\phi}) \left[s^\alpha(\boldsymbol{\mu}, T) - s^\ell(\boldsymbol{\mu}, T)\right], \tag{A.10}$$

which is related to the latent heat of fusion. In terms of Eqs. (A.8), (A.9) and (A.10), the evolution equation for temperature in Eq. (A.2) becomes

$$\begin{aligned} \frac{\partial T}{\partial t} = \frac{1}{c_p(\boldsymbol{\phi}, \boldsymbol{\mu}, T)} \Bigg[&\nabla \cdot k(\boldsymbol{\phi}, T) \nabla T - T \sum_{\alpha}^{N} g'_\alpha(\boldsymbol{\phi}) \left[s^\alpha(\boldsymbol{\mu}, T) - s^\ell(\boldsymbol{\mu}, T)\right] \frac{\partial \phi_\alpha}{\partial t} \\ &- T \sum_{i}^{n-1} \left(\sum_{\alpha}^{N} g_\alpha(\boldsymbol{\phi}) \frac{\partial s^\alpha(\boldsymbol{\mu}, T)}{\partial \mu_i} + \left[1 - \sum_{\alpha}^{N} g_\alpha(\boldsymbol{\phi})\right] \frac{\partial s^\ell(\boldsymbol{\mu}, T)}{\partial \mu_i} \right) \frac{\partial \mu_i}{\partial t} \Bigg] \end{aligned} \tag{A.11}$$

The concept of anti-trapping current can also be considered for a temperature field as latent heat can also be trapped. This effect was initially studied by Almgren for solidification of a pure material [61] where different thermal diffusivities were considered. However, since the time scale of solute diffusion is orders of magnitude slower than temperature diffusion, this effect will be neglected for alloy solidification.[1]

[1]If we need to worry about spurious kinetics for temperature, we may also require a different interpolation function for its evolution, analogous to the work of Rameriez et al. and co-workers [11].

The order parameter equation remains the same, aside from including some temperature dependence, i.e.,

$$\frac{\partial \phi_\alpha}{\partial t} = M_{\phi_\alpha} \left[\sigma_\alpha \nabla^2 \phi_\alpha - H_\alpha f'_{\rm DW}(\phi_\alpha) - \sum_{\beta \neq \alpha} \Psi'(\phi_\alpha, \phi_\beta) \right. \\ \left. - g'_\alpha(\boldsymbol{\phi}) \left[\omega^\alpha(\boldsymbol{\mu}, T) - \omega^\ell(\boldsymbol{\mu}, T) \right] \right] \tag{A.12}$$

The evolution of chemical potential is once again derived from the mass transport equation in Eq. (5.2), where the left-hand side is once again derived from the concentration field $c_i(\boldsymbol{\phi}, \boldsymbol{\mu}, T)$ given by Eq. (4.6) with $c_i^\vartheta(\boldsymbol{\mu}) \to c_i^\vartheta(\boldsymbol{\mu}, T)$ $(\vartheta = \alpha, l)$, i.e. the auxiliary concentrations explicitly depend on T, unlike the isothermal case discussed in the main text. Applying the chain rule to $\partial c_i(\boldsymbol{\phi}, \boldsymbol{\mu}, T)/\partial t$ gives

$$\frac{\partial \mu_i}{\partial t} = \frac{1}{\chi(\boldsymbol{\phi}, \mu_i, T)} \left\{ \nabla \cdot \left[\sum_j M_{ij}(\boldsymbol{\phi}, T) \nabla \mu_i \right. \right. \\ \left. + \sum_\alpha a(\boldsymbol{\phi}) W \left[c_i^\ell(\boldsymbol{\mu}, T) - c_i^\alpha(\boldsymbol{\mu}, T) \right] \frac{\partial \phi_\alpha}{\partial t} \frac{\nabla \phi_\alpha}{|\nabla \phi_\alpha|} \right] \\ - \sum_\alpha h'_\alpha(\boldsymbol{\phi}) \left[c_i^\alpha(\boldsymbol{\mu}, T) - c_i^\ell(\boldsymbol{\mu}, T) \right] \frac{\partial \phi_\alpha}{\partial t} \\ \left. - \left(\sum_\alpha^N h_\alpha(\boldsymbol{\phi}) \frac{\partial c_i^\alpha(\boldsymbol{\mu}, T)}{\partial T} + \left[1 - \sum_\alpha^N h_\alpha(\boldsymbol{\phi}) \right] \frac{\partial c_i^\ell(\boldsymbol{\mu}, T)}{\partial T} \right) \frac{\partial T}{\partial t} \right\}, \tag{A.13}$$

where we have included anti-trapping terms and swapped the interpolation functions $g_\alpha(\boldsymbol{\phi}) \to h_\alpha(\boldsymbol{\phi})$ in order to add degrees of freedom that can be used to map the model onto an appropriate sharp interface limit.

APPENDIX B

Asymptotic Analysis of the Grand Potential Phase Field Equations

THIS APPENDIX PERFORMS A MATCHED asymptotic analysis of the grand potential phase field model expressed in the basic form of Eqs. (6.6) and (6.9). We only consider a solid(α)–liquid(l) interface of a binary alloy. Generalization to multiple components is straightforward and left to the reader. We follow the mathematical manipulations of Provatas and Elder [38], with some changes due to the replacement of a special class of Helmholtz free energies by a grand potential energies. *In this chapter, we simplify notation by denoting the width of the solid-liquid interface by* W_ϕ*, the capillary length* d_o*, the kinetic time scale of the phase field equation by* τ*, and the coupling constant by* $\lambda = 1/H$*, where* H *is the nucleation barrier. Other model details are as defined and denoted as in Chapter 6. We dropped the "hat" symbol from* λ *for simplicity*

The phase field model is analyzed to $\mathcal{O}(\epsilon^2)$, where $\epsilon = W_\phi/d_o \ll 1$ is a perturbation parameter in terms of which we analyze progressively high order solutions of the phase field equations. Using this parameter may seem strange when working with *diffuse* interfaces. However, *it turns out that the relevant expansion parameter in phase field theories is* ϵ *multiplied by an effective driving force*, which is typically very small at slow cooling rates, even when ϵ becomes large. Hence, our parameter relationship derived herein will continue to hold at low cooling rates (small interface velocities) even for $\epsilon \sim \mathcal{O}(1)$ or larger. At high cooling

DOI: 10.1201/9781003204312-B

rates, but still in the limit of $\epsilon \ll 1$, the perturbative solutions of the phase field model will afford us a view into the model's continuous growth regime [53, 54].

B.1 LENGTH AND TIME SCALES

The small parameter $\epsilon = W_\phi/d_o \ll 1$ can be considered a ratio of the *inner* and *outer* length scales of the model. Alternatively, $\epsilon \equiv W_\phi/(D_L/v_s) \ll 1$, where $v_s = D_L/d_o$ is the characteristic speed of diffusion across the capillary length scale set by d_o [10] (D_L is the liquid diffusion coefficient). We also assume that $W_\phi \kappa \ll 1$ where κ is the local interface curvature. A convenient characteristic time scale for the model is the ratio $t_c = D_L/v_s^2$. A summary of the length and time scales that will guide the following analysis is summarized as follows:

$$\begin{aligned} \text{inner region} : x &\ll W_\phi \\ \text{outer region} : x &\gg D_L/v_s = d_o \\ \text{characteristic time} : t_c &= D_L/v_s^2 = d_o/v_s \\ \text{expansion parameter} : \epsilon &= W_\phi v_s/D_L = W_\phi/d_o \ll 1 \\ \text{crystal curvature} : W_\phi\, \kappa &\sim \epsilon \end{aligned} \tag{B.1}$$

It will turn out that d_o is proportional to the interface width W_ϕ and the nucleation barrier $1/\lambda$, where the units of $[\lambda] = \Omega/RT$, where Ω is the molar volume. It is thus convenient to write λ as

$$\lambda = \frac{W_\phi}{\alpha\, d_o} = \frac{\epsilon}{\alpha} \tag{B.2}$$

where $\alpha \propto RT/\Omega$, where the proportionality constant can be deduced later. The constant α can alternatively be subsumed into the scaling of the grand potential energy. Regardless, it will drop out of the analysis at each step.

B.2 PHASE FIELD EQUATIONS WRITTEN IN PERTURBATION VARIABLES

In terms of ϵ, the phase field equations in Eqs. (6.6) and (6.9) are given by

$$\tau \frac{\partial \phi}{\partial t} = W_\phi^2 \nabla^2 \phi - g'(\phi) - \frac{\epsilon}{\alpha} \left[\omega^\alpha(\mu) - \omega^\ell(\mu)\right] P'(\phi) \tag{B.3}$$

for the phase field, and

$$\chi(\phi,\mu)\frac{\partial\mu}{\partial t} = \nabla\cdot(D_L q(\phi,\mu)\nabla\mu) + \nabla\cdot\left(a_t(\phi)W_\phi\left[c^\alpha(\mu)-c^\ell(\mu)\right]\frac{\nabla\phi}{|\nabla\phi|}\frac{\partial\phi}{\partial t}\right) - h'(\phi)\left[c^\alpha(\mu)-c^\ell(\mu)\right]\frac{\partial\phi}{\partial t} \tag{B.4}$$

for the chemical potential field. Here we write $q(\phi,\mu)$ as

$$\frac{M}{D_L} \equiv q(\phi,\mu) = \tilde{q}(\phi)\chi(\phi,\mu) \tag{B.5}$$

which is a single-component generalization of Eq. (9.17), where $\tilde{q}(\phi)$ is some continuous interpolation function that varies from 0 in the solid to 1 in the liquid and the susceptibility $\chi(\phi,\mu)$. Here, $q(\phi,\mu)$ is formally expressed such as to denoted dependence on both ϕ and μ (via the susceptibility). The constant D_L is the diffusion of the liquid, and the function $h(\phi)$ retains its original meaning from Eq. (6.2) for one order parameter. In a slight change of symbols, $g_\alpha(\phi)$ in Eq. (6.9) is replaced by $P(\phi)$, which varies continuously between $\phi = 0(L)$ to $\phi = 1(\alpha)$. Similarly, the notation $f_{\mathrm{DW}}(\phi)$ for the double well potential is replaced here by $g(\phi)$. Finally, $a_t(\phi)$ here denotes is the anti-trapping flux introduced in Chapter 6, which for *simplicity will be considered to have a negative sign throughout this appendix, with a reminder that it goes back positive in Eq. (6.6), and equations derived form it.*

B.2.1 Convenient Notations and Definitions

To further simplify the notation, we will hereafter reserve the variable $\Delta\omega$ to denote the potential driving force term in the phase field equation,

$$\frac{1}{\alpha}\left[\omega^\alpha(\mu) - \omega^\ell(\mu)\right] \equiv \Delta\omega(\mu), \tag{B.6}$$

while $\Delta\tilde{c}$ represents solid (α) minus liquid (l) equilibrium concentration functions evaluated at μ,

$$\left[c^\alpha(\mu) - c^l(\mu)\right] \equiv \Delta\tilde{c}(\mu), \tag{B.7}$$

and

$$\left[c^l(\mu) - c^\alpha(\mu)\right] \equiv \Delta c(\mu) \tag{B.8}$$

When examining the outer solutions it will be instructive to convert Eq. (B.4) back to its concentration form,

$$\frac{\partial c}{\partial t} = \nabla \cdot (D_L q(\phi,\mu)\nabla\mu) + \nabla \cdot \left(a_t(\phi) W_\phi \left[c^\alpha(\mu) - c^\ell(\mu) \right] \frac{\nabla\phi}{|\nabla\phi|} \frac{\partial \phi}{\partial t} \right) \tag{B.9}$$

B.3 FIELD EXPANSIONS AND MATCHING CONDITIONS OF OUTER/INNER SOLUTIONS

The *outer* regions solutions for the phase field, for example, are denoted by ϕ^o, while in the inner region they are denoted by ϕ^{in}. Similarly for the chemical potential or the concentration fields. These solutions are assumed to expandable in an asymptotic series as

$$\begin{aligned}
\phi^o &= \phi_0^o + \epsilon\phi_1^o + \epsilon^2\phi_2^o + \cdots \\
\mu^o &= \mu_0^o + \epsilon\mu_1^o + \epsilon^2\mu_2^o + \cdots \\
c^o &= c_0^o + \epsilon c_1^o + \epsilon^2 c_2^o + \cdots ,
\end{aligned} \tag{B.10}$$

while the *inner* solutions are denoted

$$\begin{aligned}
\phi^{\text{in}} &= \phi_0^{\text{in}} + \epsilon\phi_1^{\text{in}} + \epsilon^2\phi_o^{\text{in}} + \cdots \\
\mu^{\text{in}} &= \mu_0^{\text{in}} + \epsilon\mu_1^{\text{in}} + \epsilon^2\mu_2^{\text{in}} + \cdots \\
c^{\text{in}} &= c_0^{\text{in}} + \epsilon c_1^{\text{in}} + \epsilon^2 c_2^{\text{in}} + \cdots \\
v_n &= v_0 + \epsilon v_1 + \epsilon^2 v_2 + \cdots
\end{aligned} \tag{B.11}$$

where v_n is the normal velocity, similarly expanded in powers of ϵ.

Consider a one dimensional co-ordinate (u) that runs transverse to the interface (more on this below). It is instructive to scale space (u) in the phase field equations by $\xi \equiv u/W_\phi$ to examine the *inner* solutions, and by $\eta = u/(D_L/v_s)$ is used to examine the *outer* solutions [61]. The inner and outer solutions are matched by comparing the inner solutions in the limit of $\xi \equiv u/W_\phi \to \infty$ with the outer solutions in the limit in the limit $\eta = u/(D_L/v_s) \to 0$ [61]. This leads to the matching conditions that follow between the inner and outer solutions.

For the chemical potential μ:

$$\begin{aligned}
\lim_{\xi\to\pm\infty} \mu_0^{\text{in}}(\xi) &= \mu_0^o(0^\pm) \\
\lim_{\xi\to\pm\infty} \mu_1^{\text{in}}(\xi) &= = \mu_1^o(0^\pm) + \frac{\partial\mu_0^o(0^\pm)}{\partial\eta}\xi \\
\lim_{\xi\to\pm\infty} \frac{\partial\mu_2^{\text{in}}(\xi)}{\partial\xi} &= = \frac{\partial\mu_1^o(0^\pm)}{\partial\eta} + \frac{\partial^2\mu_0^o(0^\pm)}{\partial\eta^2}\xi
\end{aligned} \tag{B.12}$$

and a similar set of matching conditions also holds for concentration c. For the phase field ϕ:

$$\begin{aligned}
\lim_{\xi\to-\infty} \phi_0^{\text{in}}(\xi) = \phi_s = \lim_{\eta\to0^+} \phi_0^o(\eta) \\
\lim_{\xi\to\infty} \phi_0^{\text{in}}(\xi) = \phi_L = \lim_{\eta\to0^-} \phi_0^o(\eta) \\
\lim_{\xi\to\pm\infty} \phi_j^{\text{in}}(\xi) = 0, \ \forall\ j = 1,2,3,\cdots \\
\phi_j^o(\eta) = 0, \ \forall\ j = 1,2,3,\cdots
\end{aligned} \tag{B.13}$$

where ϕ_s and ϕ_L denote the steady state order parameter of the bulk solid and liquid. Velocity is dependent only of the arclength s and does not require matching in the transverse co-ordinate.

B.4 OUTER EQUATIONS SATISFIED BY PHASE FIELD EQUATIONS

Re-scaling the phase field equations according to $\bar{x} = v_s x/D_L$, $\bar{y} = v_s y/D_L$ and $\bar{t} = t/(D_L/v_s^2)$ leads to the following dimensionless version of Eq. (B.3) and Eq. (B.9),

$$\bar{D}\epsilon^2\frac{\partial\phi}{\partial\bar{t}} = \epsilon^2\bar{\nabla}^2\phi - \frac{dg}{d\phi} - \epsilon\frac{\partial\Delta\omega}{\partial\phi} \tag{B.14}$$

$$\frac{\partial c}{\partial\bar{t}} = \bar{\nabla}\cdot\left\{q(\phi,c)\bar{\nabla}\mu\right\} \tag{B.15}$$

where $\bar{D} = D_L\tau/W_\phi^2$ and $\bar{\nabla}$ denotes gradients with respect to $(\bar{x},\bar{y})$.

Substituting Eqs. (B.10) into Eqs. (B.14) and (B.15) and expanding all terms up to order ϵ^2 gives many terms. Collecting terms of similar order ϵ into separate equations gives,

$$\mathcal{O}(1){:}\ g'(\phi_0^o) = 0 \tag{B.16}$$

$$\mathcal{O}(\epsilon){:}\ \Delta\omega_{,\phi}(\phi_0^o, \mu_0^o) + g''(\phi_0^o)\phi_1^o = 0 \tag{B.17}$$

$$\mathcal{O}(\epsilon^2){:}\ \bar{D}\frac{\partial \phi_0^o}{\partial \bar{t}} - \bar{\nabla}^2\phi_0^o + \Big(\Delta\omega_{,\phi\phi}(\phi_0^o, \mu_0^o)\phi_1^o + \Delta\omega_{,\phi\mu}(\phi_0^o, \mu_0^o)\mu_1^o \\ + g''(\phi_0^o)\phi_2^o + g'''(\phi_0^o)\,(\phi_1^o)^2/2\Big) = 0 \tag{B.18}$$

where $\Delta\omega_{,\phi}(\phi_0^o, \mu_0^o) \equiv \Delta\omega(\mu_0^o)P'(\phi_0^o)$. The solutions of Eq. (B.16) define the bulk values and/or minima of the double well potential function $g(\phi)$, i.e., $\phi_0^o = \phi_s$ or $\phi_0^o = \phi_L$, where ϕ_L is typically 0. Since, by construction, $P'(\phi_s) = P'(\phi_L) = 0$, Eq. (B.17) implies that $\phi_1^o = 0$. The above conditions then also trivially imply that $\phi_2^o = 0$.

Expanding the concentration equation Eq. (B.15) to second order gives the same diffusion equation to all orders in ϵ, namely,

$$\frac{\partial c_j^o}{\partial \bar{t}} = \bar{\nabla} \cdot \left\{ q(\phi_0^o, c_0^o)\bar{\nabla}\mu_j^o \right\} \tag{B.19}$$

Putting this back in dimensional units (using the scaling for $\bar{t}$ and $(\bar{x}, \bar{y})$ given at the beginning of this section) and using the fact that $\tilde{q}(\phi_0^o = \phi_L) = 1$ and $\tilde{q}(\phi_0^o = \phi_s) = D_s/D_L \equiv \xi$ gives

$$\frac{\partial c_j^o}{\partial t} = \nabla \cdot \left\{ M_{L,s}\nabla\mu_j^o \right\}, \quad \forall\, j = 0, 1, 2, \cdots \tag{B.20}$$

Thus, the usual law of diffusion holds within the bulks of either phase, far from the interface.

B.5 INNER EQUATIONS SATISFIED BY PHASE FIELD EQUATIONS

Eqs. (B.3) and Eq. (B.4) are next studied in the *inner domain*, close enough to the interface to be able to resolve structure of the fields there. The first order of business is to express the phase field equations in *curvilinear* co-ordinates, denoted (u, s) and illustrated in Figure B.1. In this coordinate system, distances are measured with respect to a co-ordinate system which is anchored to the solid-liquid interface, where u measures the distance normal to the interface to a point (x, y) and s measures the arclength from a reference position on the interface to the position on the interface coinciding with the normal from which u is measured.

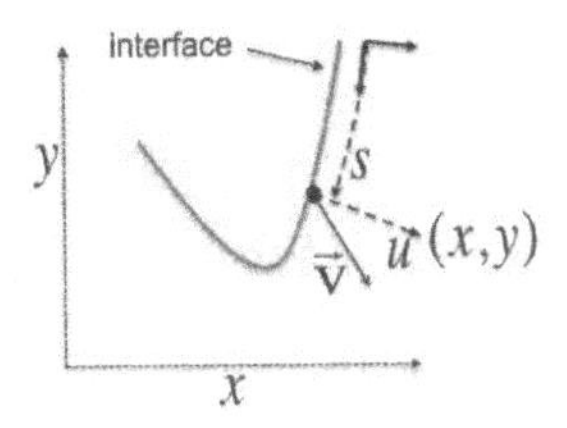

Figure B.1 Schematic of the (u, s) co-ordinates in an orthogonal co-ordinate system anchored onto the interface. The co-ordinate u measures distances normal from the interface while s measure the arclength long the interface. The vector $\vec{v}$ is the velocity of the interface at the point indicated by the dot, which is situated at co-ordinates $(0, s)$.

In terms of these co-ordinates, it can be shown that Laplacian operator transform according to

$$\nabla^2 \rightarrow \frac{\partial^2}{\partial^2 u} + \frac{\kappa}{(1+u\kappa)}\frac{\partial}{\partial u} + \frac{1}{(1+u\kappa)^2}\frac{\partial^2}{\partial s^2} - \frac{u}{(1+u\kappa)^3}\frac{\partial \kappa}{\partial s}\frac{\partial}{\partial s}, \quad \text{(B.21)}$$

where κ is the local interface curvature at the point $(0, s)$ (e.g. the dot in Figure B.1). When the diffusion coefficient is not a constant, the Laplacian is replaced by a "sandwiched" operator $\nabla \cdot (q\nabla)$, which transforms into $(u.s)$ co-ordinates according to

$$\nabla\cdot(q\nabla) \rightarrow \frac{\partial}{\partial u}\left(q\frac{\partial}{\partial u}\right) + \frac{q\kappa}{(1+u\kappa)}\frac{\partial}{\partial u} + \frac{1}{(1+u\kappa)^2}\frac{\partial}{\partial s}\left(q\frac{\partial}{\partial s}\right) - \frac{uq}{(1+u\kappa)^3}\frac{\partial \kappa}{\partial s}\frac{\partial}{\partial s} \quad \text{(B.22)}$$

Time derivatives also transform in the (u, s) co-ordinate system, according to the rule

$$\frac{\partial}{\partial t} \rightarrow \frac{\partial}{\partial t} - v_n\frac{\partial}{\partial u} + s_{,t}\frac{\partial}{\partial s} \quad \text{(B.23)}$$

The derivation of and further details about Eq. (B.21), Eq. (B.22) and Eq. (B.23) can be found in Ref. [38] and will simply be used in this appendix to derive the remainder of the matched asymptotic analysis of the phase field equations, Eqs. (B.3) and (B.4).

To proceed further, we re-scale Eqs. (B.3) and (B.4) with the *inner* space and time variables, $\xi = u/W_\phi$, $\bar{t} = t/(D_L/v_s^2)$, respectively, and arclength is re-scaled according to $\sigma = s/(D_L/v_s)$. The normal interface velocity is also re-scaled by $\bar{v}_n = v_n/v_s$ ($v_s = D_L/d_o$), while curvature is scaled by $\bar{\kappa} = W_\phi/\epsilon$. The liquid diffusion is scaled by $\bar{D} \equiv D_L\tau/W_\phi^2$.

Once Eqs. (B.3) and (B.4) are re-scaled with the above inner variables, the expansion Eq. (B.11) is substituted into the re-scaled phase field equations, and all non-linear terms are expanded to ϵ^2 order. The results are collected into distinct equations for ϕ and μ, each defined by different order of ϵ. For details of these expansions[1], the reader is referred to the appendix of Ref. [38]. The results of these expansions come out as follows.

B.5.1 Phase Field Equation

The perturbation expansion of the order parameter equation scaled with the aforementioned inner variables becomes,

$$\mathcal{O}(1):\frac{\partial^2 \phi_0^{\text{in}}}{\partial \xi^2} - g'(\phi_0^{\text{in}}) = 0 \tag{B.24}$$

$$\mathcal{O}(\epsilon):\frac{\partial^2 \phi_1^{\text{in}}}{\partial \xi^2} - g''(\phi_0^{\text{in}})\phi_1^{\text{in}} = -(\bar{D}\bar{v}_0 + \bar{\kappa})\frac{\partial \phi_0^{\text{in}}}{\partial \xi} + \Delta\omega(\mu_0^{\text{in}})P'(\phi_0^{\text{in}}) \tag{B.25}$$

$$\begin{aligned}\mathcal{O}(\epsilon^2):\frac{\partial^2 \phi_2^{\text{in}}}{\partial \xi^2} - g''(\phi_0^{\text{in}})\phi_2^{\text{in}} &= \bar{D}\frac{\partial \phi_0^{\text{in}}}{\partial \bar{t}} - \frac{\partial^2 \phi_0^{\text{in}}}{\partial u^2} - (\bar{D}\bar{v}_0 + \bar{\kappa})\frac{\partial \phi_1^{\text{in}}}{\partial \xi} \\ &- (\bar{D}\bar{v}_1 - \xi\bar{\kappa}^2)\frac{\partial \phi_0^{\text{in}}}{\partial \xi} + \Delta\omega(\mu_0^{\text{in}})P''(\phi_0^{\text{in}})\phi_1^{\text{in}} + g'''(\phi_0^{\text{in}})\frac{(\phi_1^{\text{in}})^2}{2} \\ &- \frac{\Delta\tilde{c}(\mu_0^{\text{in}})}{\alpha}P'(\phi_0^{\text{in}})\mu_1^{\text{in}}\end{aligned} \tag{B.26}$$

where $\Delta\omega$ and $\Delta\tilde{c}$ are defined by Eq. (B.6) and Eq. (B.8), respectively.

B.5.2 Chemical Potential Equation

It will prove convenient to expand Eq. (B.9) explicitly, relating c to μ separately through a constitutive relationship later. The expansion of

[1] A tricky part of these expansions are the terms of the form $\nabla \cdot (\vec{J}_\mu + \vec{J}_a)$, which appear in the chemical concentration equation. This is covered in detail in Section A.9.3 of Ref. [38] with the aid of Eq. (C17).

Eq. (B.9) yields

$$\mathcal{O}(1): \ \frac{\partial}{\partial \xi}\left(q(\phi_0^{\rm in},\mu_0^{\rm in})\frac{\partial \mu_0^{\rm in}}{\partial \xi}\right) = 0 \tag{B.27}$$

$$\mathcal{O}(\epsilon): \ \frac{\partial}{\partial \xi}\left(q(\phi_0^{\rm in},\mu_0^{\rm in})\frac{\partial \mu_1^{\rm in}}{\partial \xi}\right) \tag{B.28}$$

$$= -\frac{\partial}{\partial \xi}\left(\left\{q_{,\phi}(\phi_0^{\rm in},\mu_0^{\rm in})\phi_1^{\rm in} + q_{,\mu}(\phi_0^{\rm in},\mu_0^{\rm in})\mu_1^{\rm in}\right\}\frac{\partial \mu_0^{\rm in}}{\partial \xi}\right)$$

$$-\bar{v}_0\frac{\partial c_0^{\rm in}}{\partial \xi} - \bar{\kappa}\, q(\phi_0^{\rm in},\mu_0^{\rm in})\frac{\partial \mu_0^{\rm in}}{\partial \xi} - \frac{\partial}{\partial \xi}\left(a_t(\phi_0^{\rm in})\Delta\tilde{c}(\mu_0^{\rm in})\bar{v}_0\frac{\partial \phi_0^{\rm in}}{\partial \xi}\right)$$

$$\mathcal{O}(\epsilon^2): \ \frac{\partial}{\partial \xi}\left(q(\phi_0^{\rm in},\mu_0^{\rm in})\frac{\partial \mu_2^{\rm in}}{\partial \xi}\right) = -\bar{v}_1\frac{\partial c_0^{\rm in}}{\partial \xi} - \bar{v}_0\frac{\partial c_1^{\rm in}}{\partial \xi} \tag{B.29}$$

$$-\bar{\kappa}\, q(\phi_0^{\rm in},\mu_0^{\rm in})\frac{\partial \mu_1^{\rm in}}{\partial \xi} - \frac{\partial}{\partial \sigma}\left(q(\phi_0^{\rm in},\mu_0^{\rm in})\frac{\partial \mu_0^{\rm in}}{\partial \sigma}\right)$$

$$-\frac{\partial}{\partial \xi}\left(q_{,\phi}(\phi_0^{\rm in},c_0^{\rm in})\phi_1^{\rm in}\frac{\partial \mu_1^{\rm in}}{\partial \xi}\right) - \frac{\partial}{\partial \xi}\left(q_{,\mu}(\phi_0^{\rm in},\mu_0^{\rm in})\mu_1^{\rm in}\frac{\partial \mu_1^{\rm in}}{\partial \xi}\right)$$

$$-\frac{\partial}{\partial_\xi}\left(a_t(\phi_0^{\rm in})\bar{v}_0\Delta\tilde{c}(\mu_0^{\rm in})\frac{\partial \phi_1^{\rm in}}{\partial \xi}\right) - \bar{\kappa}\, a_t(\phi_0^{\rm in})\bar{v}_0\Delta\tilde{c}(\mu_0^{\rm in})\frac{\partial \phi_0^{\rm in}}{\partial \xi}$$

$$-\frac{\partial}{\partial \xi}\left\{\left(a_t(\phi_0^{\rm in})\bar{v}_1\Delta\tilde{c}(\mu_0^{\rm in}) + a_t(\phi_0^{\rm in})\bar{v}_0\Delta\tilde{c}_{,\mu}(\mu_0^{\rm in})\mu_1^{\rm in} + a_t'(\phi_0^{\rm in})\bar{v}_0\Delta\tilde{c}(\mu_0^{\rm in})\phi_1^{\rm in}\right)\right.$$

$$\left.\frac{\partial \phi_0^{\rm in}}{\partial \xi}\right\}$$

where primes denotes differentiation with respect ϕ and subscripts following a comma denote differentiation with respect to the subscripted variable. In the above, it was assume that $c_o^{\rm in}$ and $\phi_o^{\rm in}$ do not depend explicitly on time, and $\sigma_{,\bar{t}}$ was assumed to be negligible.

B.5.3 Constitutive Relation between c and μ

The above perturbation expansion is closed by recalling the constitutive relation between c and μ given by Eq. (6.2), which we denote symbolically as $c = \boldsymbol{\mu}^{-1}(\phi,\mu)$. Substituting the inner expansions for c, ϕ, μ from Eq. (B.11) into both sides of c=$\boldsymbol{\mu}^{-1}(\phi,\mu)$, expanding both sides to order ϵ and comparing terms that emerge at different powers of ϵ separately gives,

$$c_0^{\rm in} = h(\phi_0^{\rm in})c^\alpha(\mu_0^{\rm in}) + (1 - h(\phi_0^{\rm in}))c^l(\mu_0^{\rm in}) \tag{B.30}$$

$$c_1^{\rm in} = \left[c^\alpha(\mu_0^{\rm in}) - c^l(\mu_0^{\rm in})\right]h'(\phi_0^{\rm in})\phi_1^{\rm in} + \chi(\phi_0^{\rm in},\mu_0^{\rm in})\mu_1^{\rm in} \tag{B.31}$$

B.6 ANALYSIS OF INNER EQUATIONS AND MATCHING TO THEIR OUTER FIELDS

The next step is to solve the various equations generated at different orders of ϵ by the perturbation expansion for ϕ and μ, and use the matching conditions in Eqs. (B.12) and (B.13) to *tune the parameters of the phase field equations* so that the outer solutions satisfy the sharp interface model of alloy solidification.

B.6.1 $\mathcal{O}(1)$ Phase Field Equation (B.24)

Equation (B.24) can be solved analytically [38]. As an example, for the classical double-well potential $g(\phi) = -\phi^2/2 + \phi^4/4$,

$$\phi_0^{\text{in}} = -\tanh\left(\frac{\xi}{\sqrt{2}}\right) \tag{B.32}$$

In the example considered above, the far field values of the hyperbolic tangent function are $\phi_s = 1$ an $\phi_L = -1$, which define the minima of $g(\phi)$. Transforming $g(\phi)$ by $\phi \rightarrow 2\phi - 1$, where $0 < \phi < 1$ (as assumed here) gives $\phi_0^{\text{in}} = -(1/2)\left(\tanh\left(\xi/\sqrt{2}\right) - 1\right)$.

B.6.2 $\mathcal{O}(1)$ Diffusion Equation (B.27)

Integrating Eq. (B.27) gives,

$$\frac{\partial \mu_0^{\text{in}}}{\partial \xi} = \frac{B}{q(\phi_0^{\text{in}}, \mu_0^{\text{in}})} \tag{B.33}$$

where B is an integration constant that may depend of the arclength σ. Integrating Eq. (B.33) once more gives,

$$\mu_0^{\text{in}} = \mu_{\text{E}}(\sigma) + B\int_{-\infty}^{\xi} \frac{d\xi}{q(\phi_0^{\text{in}}, \mu_0^{\text{in}})} \tag{B.34}$$

where $\mu_{\text{E}}(\sigma)$ is a second integration constant, also dependent on the [scaled] arclength σ since integration is with respect to ξ. Since $q(\phi_0^{\text{in}}, \mu_0^{\text{in}})$ becomes a constant in the liquid, i.e. as $\xi \rightarrow \infty$, the limit $\lim_{\xi\rightarrow\pm\infty} \mu_0^{\text{in}}(\xi) = \mu_{\text{E}} + \lim_{\xi\rightarrow\infty} \int_{-\infty}^{\xi} 1/q(\phi_0^{\text{in}}, \mu_0^{\text{in}}) d\xi$ diverges unless $B = 0$. Taking these considerations into account gives[2]

$$\mu_0^{\text{in}} \equiv \mu_0^{\text{in}}(\sigma) \tag{B.35}$$

[2] μ_0^{in} can also depend on interface speed, hereafter tacitly assumed, but suppressed for ease of notation.

The first of matching conditions in Eq. (B.12) requires that

$$\lim_{\xi\to\pm\infty} \mu_0^{\text{in}} \equiv \mu_0^{\text{in}}(\sigma) = \lim_{\eta\to 0^\pm} \mu_0^o(\eta) \equiv \mu_0^o(0^\pm), \tag{B.36}$$

where $\mu_0^o(0^\pm)$ is the lowest order chemical potential of the outer domain projected onto the solid/liquid interface. Thus, according to Eq. (B.36), $\mu_0^o(0^\pm)$ is a the same value on both sides of the interface, and depends only on the local arclength.

It is noteworthy that Eq. (B.36) implies that

$$\frac{\partial \bar{f}_{\text{AB}}^{\text{mix}}(\phi_s, c_s)}{\partial c} = \mu_0^{\text{in}}(\sigma) = \mu_0^o(0^\pm) \tag{B.37}$$

$$\frac{\partial \bar{f}_{\text{AB}}^{\text{mix}}(\phi_L, c_L)}{\partial c} = \mu_0^{\text{in}}(\sigma) = \mu_0^o(0^\pm) \tag{B.38}$$

where $\bar{f}_{AB}^{\text{mix}}$ is the alloy free energy. If $\bar{f}_{AB}^{\text{mix}}$ is invertible, Eqs. (B.37) and (B.38) define

$$c_L \equiv c^l(\mu_0^o(0^\pm)), \quad c_s \equiv c^\alpha(\mu_0^o(0^\pm)) \tag{B.39}$$

as the lowest order bulk concentration values at the solid and liquid sides of the interface, respectively; these are the far-field limits of Eq. (B.30). Next section will show that c_s and c_L defined by Eqs. (B.37) and (B.38) implicitly depend on local interface curvature and speed via the Gibbs-Thomson correction.

B.6.3 $\mathcal{O}(\epsilon)$ Phase Field Equation (B.25)

Equation (B.25) is simplified by first multiplied by $\partial\phi_0^{\text{in}}/d\xi$ and integrated from $\xi \to -\infty$ to ∞, giving

$$\int_{-\infty}^{\infty} \frac{\partial \phi_0^{\text{in}}}{\partial \xi} \mathcal{L}(\phi_1^{\text{in}}) d\xi = -(\bar{D}\bar{v}_0 + \bar{\kappa}) \int_{-\infty}^{\infty} \left(\frac{\partial \phi_0^{\text{in}}}{\partial \xi}\right)^2 d\xi + \Delta\omega(\mu_0^{\text{in}}) \int_{-\infty}^{\infty} P'(\phi_0^{\text{in}}) \frac{\partial \phi_0^{\text{in}}}{\partial \xi} d\xi \tag{B.40}$$

where $\mathcal{L} \equiv \partial_{\xi\xi} - g''(\phi_0^{\text{in}})$ and the double prime on $g(\phi)$ denotes a double derivative with respect to ϕ. It is straightforward to show that left-hand side of Eq. (B.40) vanishes using Eq. (B.24) [38]. Moreover, we denote the first integral on the right-hand side of Eq. (B.40) by

$$\sigma_\phi \equiv \int_{-\infty}^{\infty} \left(\frac{\partial \phi_0^{\text{in}}}{\partial \zeta}\right)^2 d\xi, \tag{B.41}$$

and assume, without loss of generality, that $0 < \phi < 1$ in our description of the order parameter, normalizing $P(\phi)$ according to

$$\int_{-\infty}^{\infty} P'(\phi_0^{\text{in}}) \frac{\partial \phi_0^{\text{in}}}{\partial \xi} d\xi = -1 \tag{B.42}$$

with these simplifications, Eq. (B.40) becomes

$$-(\bar{D}\bar{v}_0 + \bar{\kappa})\sigma_\phi - \Delta\omega(\mu_0^{\text{in}}) = 0 \tag{B.43}$$

To proceed with the calculation, we consider low undercooling, or *low supersaturation*, conditions and expand the grand potentials in $\Delta\omega(\mu)$ about the equilibrium chemical potential μ_{E}. This gives

$$\Delta\omega(\mu_0^{\text{in}}) = \Delta\omega(\mu_{\text{E}}) + \left.\frac{\partial \Delta\omega}{\partial \mu}\right|_{\mu_{\text{E}}} \left(\mu_0^{\text{in}} - \mu_{\text{E}}\right) + \cdots \tag{B.44}$$

In Eq. (B.44), $\Delta\omega(\mu_{\text{E}}) = 0$ by definition of μ_{E}. It is noted that differentiating Eq. (B.6) gives

$$\frac{\partial \Delta\omega(\mu_0^{\text{in}})}{\partial \mu} = \frac{1}{\alpha}\left(c^l(\mu_0^{\text{in}}) - c^\alpha(\mu_0^{\text{in}})\right) = \frac{1}{\alpha}(c_L - c_s) = \frac{\Delta c}{\alpha} \tag{B.45}$$

Thus, by the same reasoning,

$$\frac{\partial \Delta\omega(\mu_{\text{E}})}{\partial \mu} = \frac{1}{\alpha}\left(c^l(\mu_{\text{E}}) - c^\alpha(\mu_{\text{E}})\right) = \frac{1}{\alpha}(c_L^{\text{eq}} - c_s^{\text{eq}}) \equiv \frac{\Delta c_F}{\alpha} \tag{B.46}$$

Substituting Eq. (B.44) into Eq. (B.43), and using Eq. (B.36), gives

$$\mu_0^o(0^\pm) = \mu_{\text{E}} - \frac{\bar{D}\alpha\sigma_\phi}{\Delta c_F}\bar{v}_0 - \frac{\sigma_\phi \alpha}{\Delta c_F}\bar{\kappa} \tag{B.47}$$

Equation (B.47) is put into dimensional form by utilizing the variable re-scalings in the second paragraph of Section B.5. Namely, speed becomes $\bar{v}_0 = (d_o/D_L)v_0$ and curvature by $\bar{\kappa} = (W_\phi/\epsilon)\kappa$. Then use the definition of the length scale $d_o = W_\phi/(\alpha\lambda)$ from Eq. (B.2), and note that $\alpha/\epsilon = w = 1/\lambda$, which finally gives,

$$\mu_0^o(0^\pm) = \mu_{\text{E}} - \left(\frac{\sigma_\phi}{\Delta c_F}\right)\left(\frac{W_\phi}{\lambda}\right)\kappa - \left(\frac{\sigma_\phi}{\Delta c_F}\right)\left(\frac{\tau}{\lambda W_\phi}\right) v_0 \tag{B.48}$$

It is noteworthy that if instead of proceeding by expanding $\Delta\omega(\mu_0^{\text{in}})$ in Eq. (B.44), we took the route of writing out $\Delta\omega(\mu_0^{\text{in}})$ explicitly using the definition $\omega = f - \mu c$, (see steps following Eq. C68 in Ref. [38]), would have arrived to the same form of the Gibb Thomson equation in Eq. (B.48) except that $\Delta c_F \to \Delta c$. The concentration jump Δc can be written as $\Delta c = \Delta c_F(1+\delta c)$ where $\Delta c_F \equiv c_L^{\text{eq}} - c_s^{\text{eq}}$ and δc is small curvature and velocity correction according to Eq. (B.37), Eq. (B.38) and Eq. (B.48). As discussed in Refs. [10,38], this correction implies that we cannot set the sharp interface kinetic coefficient β exactly to zero as it depends on the local interface conditions. This can be mitigated in certain ways [10]; for the remainder of this appendix, we will simply assume that $\delta c \sim W_\phi \kappa \sim \epsilon \ll 1$, and $(\tau/W_\phi)v_0 \ll 1$ and approximate $\Delta c \approx \Delta c_F$.

B.6.4 $\mathcal{O}(\epsilon)$ Diffusion Equation (B.29)

Equation (B.29) is greatly simplified by observing that the $\mu_0^{\text{in}}(\sigma)$ dependence vanishes as it does not depend on ξ. Integrating the surviving equation from $\xi \to -\infty$ to ξ gives,

$$q(\phi_0^{\text{in}}, \mu_0^{\text{in}})\frac{\partial \mu_1^{\text{in}}}{\partial \xi} = -\bar{v}_0\left[c_0^{\text{in}}(\xi) + a_t(\phi_0^{\text{in}})\Delta\tilde{c}(\mu_0^{\text{in}})\frac{\partial \phi_0^{\text{in}}}{\partial \xi}\right] + A \tag{B.49}$$

The integration constant A is found by considering the $\xi \to -\infty$ limit of Eq. (B.49) and by approximating the experimentally relevant limit $\tilde{q}(\phi(\xi \to -\infty)) = \tilde{q}(\phi_s) = D_s/D_L \ll 1$ (≈ 0 in practical situations). With this assumption the boundary condition

$$\lim_{\xi\to-\infty}\left(q(\phi_0^{\text{in}}, \mu_0^{\text{in}})\frac{\partial \mu_1^{\text{in}}}{\partial \xi}\right) = 0 = -\bar{v}_0 c_0^{\text{in}}(-\infty) + A \tag{B.50}$$

gives $A = \bar{v}_0\, c_s$ where by Eq. (B.30), $\lim_{\xi\to-\infty} c_0^{\text{in}}(\xi) = c^\alpha(\mu_0^{\text{in}}) \equiv c_s$ has been used. Denoting $\lim_{\xi\to\infty} c_0^{\text{in}}(\xi) = c^l(\mu_0^{\text{in}}) \equiv c_L$, and integrating Equation (B.49) once gives

$$\mu_1^{\text{in}} = -\bar{v}_0 \int_0^\xi \frac{\left[\bar{c}_0^{\text{in}}(x) - c_s\right]}{q(\phi_0^{\text{in}}, \mu_0^{\text{in}})}dx + \bar{\mu} \tag{B.51}$$

where we define

$$\bar{c}_0^{\text{in}}(\xi) \equiv c_0^{\text{in}}(\xi) + a_t(\phi_0^{\text{in}})\Delta\tilde{c}(\mu_0^{\text{in}})\frac{\partial \phi_0^{\text{in}}}{\partial \xi} \tag{B.52}$$

and $\bar{\mu}$ is an integration constant.

It is instructive to split Eq. (B.51) into two pieces, one valid for $\xi < 0$ and the other for $\xi > 0$,

$$\mu_1^{\text{in}} = -\bar{v}_0 \int_0^{\xi} \left\{ \frac{[\bar{c}_0^{\text{in}}(x) - c_s]}{q(\phi_0^{\text{in}}, \mu_0^{\text{in}})} - \frac{[c_L - c_s]}{q^+} \right\} dx - \frac{\bar{v}_0(c_L - c_s)}{q^+}\xi + \bar{\mu}, \quad \xi > 0 \tag{B.53}$$

$$\mu_1^{\text{in}} = \bar{v}_0 \int_{\xi}^{0} \frac{[\bar{c}_0^{\text{in}}(x) - c_s]}{q(\phi_0^{\text{in}}, \mu_0^{\text{in}})} dx + \bar{\mu}, \quad \xi < 0 \tag{B.54}$$

where the notation $q^+ \equiv q(\phi_L, \mu_o^{\text{in}}) = q(\phi_L, \mu_0^o(0^+))$ has been defined to simplify the notation. In terms of Eqs. (B.53) and (B.54), the far-field limits of Eq. (B.51) become,

$$\lim_{\xi \to \infty} \mu_1^{\text{in}} = \bar{v}_0 \int_0^{\infty} \left\{ \frac{\Delta c}{q^+} - \frac{[\bar{c}_0^{\text{in}}(x) - c_s]}{q(\phi_0^{\text{in}}, \mu_0^{\text{in}})} \right\} dx - \frac{\bar{v}_0 \Delta c}{q^+}\xi + \bar{\mu} \tag{B.55}$$

$$\lim_{\xi \to -\infty} \mu_1^{\text{in}} = \bar{v}_0 \int_{-\infty}^{0} \frac{[\bar{c}_0^{\text{in}}(x) - c_s]}{q(\phi_0^{\text{in}}, \mu_0^{\text{in}})} dx + \bar{\mu} \tag{B.56}$$

where Δc is defined in Eq. (B.45).

Using Eqs. (B.55) and (B.56) in the second matching condition of Eq. (B.12) gives

$$\mu_1^o(0^+) + \frac{\partial \mu_0^o(0^+)}{\partial \eta}\xi = \bar{\mu} + \bar{v}_0 F^+ - \frac{\bar{v}_0 \Delta c}{q^+}\xi \tag{B.57}$$

$$\mu_1^o(0^-) + \frac{\partial \mu_0^o(0^-)}{\partial \eta}\xi = \bar{\mu} + \bar{v}_0 F^- \tag{B.58}$$

where

$$F^+ = \int_0^{\infty} \left\{ \frac{\Delta c}{q^+} - \frac{[\bar{c}_0^{\text{in}}(x) - c_s]}{q(\phi_0^{\text{in}}, \mu_0^{\text{in}})} \right\} dx$$

$$F^- = \int_{-\infty}^{0} \frac{[\bar{c}_0^{\text{in}}(x) - c_s]}{q(\phi_0^{\text{in}}, \mu_0^{\text{in}})} dx, \tag{B.59}$$

are coined the *solute trapping* integrals F^+ and F^- here.

Subtracting Equation (B.58) from Eq. (B.57) and comparing powers of ξ^0 in the result gives,

$$\mu_1^o(0^+) - \mu_1^o(0^-) = (F^+ - F^-)\bar{v}_0 \tag{B.60}$$

Equation (B.60) can be made more illuminating by expressing $\mu^o \approx$

$\mu_0^o + \epsilon\mu_1^o + \cdots$ and replacing $\epsilon = W_\phi v_s/D_L$ and $\bar{v}_0 = v_0/v_s$. Recalling that $\mu_0^o(0^+) = \mu_0^o(0^-)$ gives

$$\mu^o(0^+) - \mu^o(0^-) = \epsilon\mu_1^o(0^+) - \epsilon\mu_1^o(0^-) = \frac{W_\phi}{D_L}(F^+ - F^-)v_0 \qquad \text{(B.61)}$$

Equation (B.60) predicts that to $\mathcal{O}(\epsilon)$, the interface thickness (W_ϕ) gives rise to a jump discontinuity in the chemical potential for moving interfaces. This effect lies at the heart of solute trapping.

Subtracting Eq. (B.58) from Eq. (B.57) and comparing powers of ξ also gives,

$$q^+\frac{\partial\mu_0^o(0^+)}{\partial\eta} = -\bar{v}_0\Delta c \qquad \text{(B.62)}$$

Eq. (B.62) is cast into dimensional units by substituting $\bar{v}_0 = v_0/v_s$ and $\eta = v_s u/D_L$, which leads to

$$D_L\chi^L\frac{\partial\mu_0^o(0^+)}{\partial u} = -v_0\Delta c \qquad \text{(B.63)}$$

where $\chi^L \equiv \chi^L(\mu_0^o(0^\pm))$. Equation (B.63) describes the flux conservation condition for *one-sided* solidification to an error of $\mathcal{O}(\epsilon)$. It assumes that $q^- \equiv q(\phi_s, \mu_o^{\text{in}}) = q(\phi_s, \mu_0^o(0^-)) \approx 0$, a reasonable condition for most alloys during solidification.

B.6.5 $\mathcal{O}(\epsilon^2)$ Phase Field Equation (B.26)

Equation (B.26) is analyzed by multiplying by $\partial\phi_0^{\text{in}}/\partial\xi$ and integrating from $\xi = -\infty$ to $\xi = \infty$. Dropping the ϕ_0^{in} terms dependent on $\bar{t}$ and σ gives

$$\begin{aligned}\int_{-\infty}^{\infty}\frac{\partial\phi_0^{\text{in}}}{\partial\xi}\mathcal{L}(\phi_2^{\text{in}})d\xi &= -\bar{D}\bar{v}_1\int_{-\infty}^{\infty}\left(\frac{\partial\phi_0^{\text{in}}}{\partial\xi}\right)^2 d\xi + \int_{-\infty}^{\infty}\frac{\partial\phi_0^{\text{in}}}{\partial\xi}\\ &\times\left\{\Delta\omega(\mu_o^{\text{in}})P''(\phi_o^{\text{in}})\phi_1^{\text{in}} - \frac{\Delta\tilde{c}(\mu_0^{\text{in}})}{\alpha}P'(\phi_o^{\text{in}})\mu_1^{\text{in}}\right\}d\xi\\ &- (\bar{D}\bar{v}_0 + \bar{\kappa})\int_{-\infty}^{\infty}\frac{\partial\phi_0^{\text{in}}}{\partial\xi}\frac{\partial\phi_1^{\text{in}}}{\partial\xi}d\xi + \bar{\kappa}^2\int_{-\infty}^{\infty}\xi\left(\frac{\partial\phi_0^{\text{in}}}{\partial\xi}\right)^2 d\xi\\ &+ \frac{1}{2}\int_{-\infty}^{\infty}\frac{\partial\phi_0^{\text{in}}}{\partial\xi}g'''(\phi_0^{\text{in}})(\phi_1^{\text{in}})^2 d\xi\end{aligned} \qquad \text{(B.64)}$$

The left-hand side of Eq. (B.64) can be made to vanish using integration by parts. The fourth integral on the right-hand side of Eq. (B.64) vanishes since the derivative of ϕ_0^{in} is symmetric or even in ξ about the origin. To address the third and fifth integrals on the right-hand side of Eq. (B.64), it is first instructive to investigate the properties of ϕ_1^{in} from Eq. (B.25),

$$\mathcal{L}(\phi_1^{\text{in}}) = -(\bar{D}\bar{v}_0 + \bar{\kappa})\frac{\partial \phi_0^{\text{in}}}{\partial \xi} + \Delta\omega(\mu_0^{\text{in}})P'(\phi_0^{\text{in}}) \tag{B.65}$$

where $\mathcal{L} \equiv \partial_{\xi\xi} - g''(\phi_0^{\text{in}})$. Choosing $g(\phi)$ to be an even function of ϕ, makes the operator $\mathcal{L}$ even in ξ since $g''(\phi_0^{\text{in}}(\xi))$ is even in ξ. Since both sides of Eq (B.65) are even in ξ, $\phi_1^{\text{in}}(\xi)$ is an even function of ξ. These considerations imply that the third integral on the right-hand side of Eq. (B.64) is zero. The last integral on on the right-hand side of Eq. (B.64) also vanishes as its integrand is odd in ξ (i.e., even function $\times$ odd function $\times$ even function). Similarly, the $P''(\phi_o^{\text{in}})$ term in the second integral of Eq. (B.64) vanishes due to symmetry.

Taking the simplifications of the previous paragraph into account reduces Eq. (B.64) to a more tractable form,

$$-\bar{D}\sigma_\phi\bar{v}_1 - \frac{\Delta\tilde{c}(\mu_0^{\text{in}})}{\alpha}\int_{-\infty}^{\infty}\frac{\partial \phi_0^{\text{in}}}{\partial \xi}P'(\phi_o^{\text{in}})\mu_1^{\text{in}}d\xi = 0 \tag{B.66}$$

Equation (B.66) will next be analyzed to yield added contributions to the Gibbs-Thomson condition to those found at order $\mathcal{O}(\epsilon)$.

Equation (B.51) is used to eliminate μ_1^{in} in Eq. (B.66), yielding

$$-\bar{D}\sigma_\phi\bar{v}_1 - \frac{\bar{v}_0\,\Delta\tilde{c}(\mu_0^{\text{in}})}{\alpha}K + \frac{\bar{\mu}\,\Delta\tilde{c}(\mu_0^{\text{in}})}{\alpha} = 0 \tag{B.67}$$

where

$$K = \int_{-\infty}^{\infty}\frac{\partial \phi_0^{\text{in}}}{\partial \xi}P'(\phi_o^{\text{in}})\left\{\int_0^{\xi}\frac{c_s - \tilde{c}_0^{\text{in}}(x)}{q(\phi_0^{\text{in}},\mu_0^{\text{in}})}dx\right\}d\xi \tag{B.68}$$

Matching the ξ^0 terms in Eq. (B.57) and Eq. (B.58) gives $\bar{\mu} = \mu_1^o(0^\pm) - \bar{v}_0F^\pm$. Substituting $\bar{\mu}$ into Eq. (B.67) leads to

$$\mu_1^o(0^\pm) = -\frac{\alpha\bar{D}\sigma_\phi}{\Delta c}\bar{v}_1 + (K + F^\pm)\,\bar{v}_0 \tag{B.69}$$

It is noteworthy that $\mu_1^o(0^\pm)$, unlike the lowest order correction $\mu_0^o(0^\pm)$, is not the same on either side of the interface, i.e. there is a chemical potential jump proportional to $\Delta F \equiv F^+ - F^-$ at the interface. This is a direct consequence of a non-zero interface (W_ϕ)[3].

[3]At small velocities, where this effect becomes negligible in experiments, while a from Eq. (B.61), it is seen that a phase field model operated at an exaggeratedly large W_ϕ for numerical efficiency will accentuate this term's significance, leading to errors.

It is instructive to re-cast $\epsilon\mu_1^o(0^\pm)$ into dimensional form by utilizing the re-scaling definitions discussed after Eq. (B.47). This gives

$$\epsilon\mu_1^o(0^\pm) = (K + F^\pm)\frac{W_\phi}{D_L}v_0 - \frac{\tau\sigma_\phi}{W_\phi\lambda\Delta c}\epsilon v_1 \tag{B.70}$$

Final form of the of Gibbs-Thomson boundary condition

Since $\mu^o \approx \mu_0^o + \epsilon\mu_1^o + \mathcal{O}(\epsilon^2)$, Eq. (B.70) and Eq. (B.48) can be added to obtain the final Gibbs-Thomson condition to an error of $\mathcal{O}(\epsilon^2)$,

$$\mu^o(0^\pm) = \mu_{\rm E} - \frac{\sigma_\phi}{\Delta c_F}\frac{W_\phi}{\lambda}\kappa - \frac{\tau\sigma_\phi}{W_\phi\lambda\Delta c_F}\left\{1 - \frac{(K + F^\pm)\,\Delta c_F\,\lambda}{\sigma_\phi\,\bar{D}}\right\}v_0 \tag{B.71}$$

where the $\epsilon v_1/\Delta c$ term from Eq. (B.70) was formally dropped since, for typical metals, it would contribute a term of order $< \mathcal{O}(\epsilon)$ to the bracketed expression in Eq. (B.71)[4]. Furthermore, from the discussion following Eq. (B.48), it is recalled that a more precise form of Eq. (B.71) should contain $\Delta c_F \to \Delta c$.

B.6.6 $\mathcal{O}(\epsilon^2)$ Diffusion Equation (B.29)

Here we extend the flux conservation condition in Eq. (B.63) to $\mathcal{O}(\epsilon^2)$ (see more details in Ref. [38]). We begin by Integrating the $\mathcal{O}(\epsilon^2)$ concentration equation (B.29) from 0 to ξ, and use Eq. (B.49) to eliminate $q(\phi_0^{\rm in}, c_0^{\rm in})\partial\mu_1^{\rm in}/\partial\xi$. This gives

$$\begin{aligned}
q(\phi_0^{\rm in}, \mu_0^{\rm in})\frac{\partial\mu_2^{\rm in}}{\partial\xi} &= -\bar{v}_1 c_0^{\rm in}(\xi) - \bar{v}_0 c_1^{\rm in}(\xi) + \bar{\kappa}\bar{v}_0\int_0^\xi \left(c_0^{\rm in}(x) - c_s\right)dx \\
&- \frac{\partial^2\mu_0^{\rm in}}{\partial\sigma^2}\int_0^\xi q(\phi_0^{\rm in}, \mu_0^{\rm in})dx \\
&- q_{,\phi}(\phi_0^{\rm in}, c_0^{\rm in})\phi_1^{\rm in}\frac{\partial\mu_1^{\rm in}}{\partial\xi} - q_{,\mu}(\phi_0^{\rm in}, \mu_0^{\rm in})\mu_1^{\rm in}\frac{\partial\mu_1^{\rm in}}{\partial\xi} - a_t(\phi_o^{\rm in})\,\bar{v}_0\,\Delta\tilde{c}(\mu_0^{\rm in})\frac{\partial\phi_1^{\rm in}}{\partial\xi} \\
&- \left(a_t(\phi_o^{\rm in})\bar{v}_1\Delta\tilde{c}(\mu_0^{\rm in}) + a_t(\phi_o^{\rm in})\bar{v}_0\Delta\tilde{c}_{,\mu}(\mu_o^{\rm in})\mu_1^{\rm in} + a_t'(\phi_o^{\rm in})\bar{v}_0\Delta\tilde{c}(\mu_0^{\rm in})\phi_1^{\rm in}\right)\frac{\partial\phi_o^{\rm in}}{\partial\xi} \\
&+ B(\sigma)
\end{aligned} \tag{B.72}$$

[4]For typical metals $(\tau/W_\phi\lambda)\epsilon v_1 \ll \mathcal{O}(\epsilon)$ and so this term can be dropped in Eq. (B.70). For the important case of $\beta = 0$ that describes solidification at low to moderate undercooling, we have $\bar{D} = a_2\lambda$, and $(\tau/W_\phi\lambda)\epsilon v_1 \sim \mathcal{O}(\epsilon^2)$. In fact, the v_1 term would not have appeared here at all if the interface velocity $v_n(s,t)$ was left unexpanded as in [10]. That is because going to the next order in the asymptotic expansion should simply reproduce the expression in the large curly brackets of Eq. (B.71) to order ϵv_1, and so on to higher order. Thus, it suffices to treat v_0 as the interface velocity, not just the lowest order velocity.

where $B(\sigma)$ is an integration constant that depends on the scaled arc-length σ.

It is not necessary to explicitly determine μ_2^{in}. Instead the $\xi \to \pm\infty$ limits (i.e. solid/liquid) limits of Eq. (B.72) only need be considered. The $\xi \to -\infty$ limit becomes

$$\begin{aligned}
\lim_{\xi\to-\infty}\left(q(\phi_0^{\text{in}},\mu_0^{\text{in}})\frac{\partial \mu_2^{\text{in}}}{\partial \xi}\right) &= q^-\frac{\partial \mu_1^o(0^-)}{\partial \eta} + q^-\frac{\partial^2 \mu_0^o(0^-)}{\partial \eta^2}\,\xi \\
&= -\bar{v}_1 c_s - \bar{v}_0 c_1^{\text{in}}(\xi\to-\infty) - \bar{\kappa}\bar{v}_0\int_{-\infty}^{0}\left(c_0^{\text{in}}(x)-c_s\right)dx \\
&+ \frac{\partial^2 \mu_0^{\text{in}}}{\partial \sigma^2}\int_{-\infty}^{0} dx\, q(\phi_0^{\text{in}},\mu_0^{\text{in}}) - q_\mu(\phi_s,\mu_0^{\text{in}})\lim_{\xi\to-\infty}\left(\mu_1^{\text{in}}\frac{\partial \mu_1^{\text{in}}}{\partial \xi}\right) \\
&+ B(\sigma)
\end{aligned} \tag{B.73}$$

The $\xi \to \infty$ limit is

$$\begin{aligned}
\lim_{\xi\to\infty}\left(q(\phi_0^{\text{in}},\mu_0^{\text{in}})\frac{\partial \mu_2^{\text{in}}}{\partial \xi}\right) &= q^+\frac{\partial \mu_1^o(0^+)}{\partial \eta} + q^+\frac{\partial^2 \mu_0^o(0^+)}{\partial \eta^2}\,\xi \\
&= -\bar{v}_1 c_L - \bar{v}_0 c_1^{\text{in}}(\xi\to\infty) + \bar{\kappa}\bar{v}_0\int_{0}^{\infty}\left(c_0^{\text{in}}(x)-c_s\right)dx \\
&+ \frac{\partial^2 \mu_0^{\text{in}}}{\partial \sigma^2}\int_{0}^{\infty} dx\, q(\phi_0^{\text{in}},\mu_0^{\text{in}}) - q_\mu(\phi_L,\mu_0^{\text{in}})\lim_{\xi\to\infty}\left(\mu_1^{\text{in}}\frac{\partial \mu_1^{\text{in}}}{\partial \xi}\right) \\
&+ B(\sigma)
\end{aligned} \tag{B.74}$$

where the last of Eq. (B.12) was used on the first lines of Eq. (B.73) and Eq. (B.74) to emphasize the connection of the limit with the outer solution. Note that while $q^- \approx 0$, we formally retain it in the following manipulations for symmetry, and drop it later.

Equation (B.73) and Eq. (B.74) are simplified by writing, respectively,

$$\begin{aligned}
\int_{-\infty}^{0} q(\phi_0^{\text{in}},\mu_0^{\text{in}})dx &= \int_{-\infty}^{0}\left(q(\phi_0^{\text{in}},\mu_0^{\text{in}}) - q^-\right)dx - q^-\,\xi \\
\int_{0}^{\infty} q(\phi_0^{\text{in}},\mu_0^{\text{in}})dx &= \int_{0}^{\infty}\left(q(\phi_0^{\text{in}},\mu_0^{\text{in}}) - q^+\right)dx + q^+\,\xi
\end{aligned} \tag{B.75}$$

We also employ the following matching conditions in both Eq. (B.73) and Eq. (B.74),

$$\begin{aligned}\lim_{\xi\to\pm\infty}\mu_1^{\rm in}(\xi) &= = \mu_1^o(0^\pm) + \frac{\partial \mu_0^o(0^\pm)}{\partial\eta}\xi \\ \lim_{\xi\to\pm\infty}\frac{\partial\mu_1^{\rm in}}{\partial\xi} &= \frac{\partial\mu_0^{\rm in}(0^\pm)}{\partial\eta} \\ \lim_{\xi\to\pm\infty}c_1^{\rm in}(\xi) &= = c_1^o(0^\pm) + \frac{\partial c_0^o(0^\pm)}{\partial\eta}\xi \end{aligned} \tag{B.76}$$

Finally, in Eq. (B.73) we make the replacement

$$\int_0^\infty \left(c_0^{\rm in} - c_s\right) dx = \int_0^\infty \left(c_0^{\rm in} - c_L\right) dx + \Delta c\,\xi \tag{B.77}$$

Substituting Eq. (B.75), Eq. (B.76) and Eq. (B.77) into Eq. (B.73) and Eq. (B.74) yields the final form $q^\pm\left(\partial\mu_1^o(0^\pm)/\partial\eta\right) + q^\pm\left(\partial^2\mu_0^o(0^\pm)/\partial\eta^2\right)\xi$. Comparing the ξ^0 terms of these resulting expressions yields the $\mathcal{O}(\epsilon^2)$ correction to the flux conservation condition

$$\begin{aligned} q^-\frac{\partial\mu_1^o(0^-)}{\partial\eta} - q^+\frac{\partial\mu_1^o(0^+)}{\partial\eta} &= \bar{v}_1\Delta c + \bar{v}_0\Delta c_1 + \bar{\kappa}\,\bar{v}_0\Delta H - \left[\partial_{\sigma\sigma}\mu_0^o(0^\pm)\right]\Delta J \\ &+ \left\{q_{,\mu}^+\,\mu_1^o(0^+)\frac{\partial\mu_0^o(0^+)}{\partial\eta} - q_{,\mu}^-\,\mu_1^o(0^-)\frac{\partial\mu_0^o(0^-)}{\partial\eta}\right\}, \end{aligned} \tag{B.78}$$

where $\Delta c_1 \equiv c_1^o(0^+) - c_1^o(0^-)$, $q_{,\mu}^+ \equiv q_{,\mu}(\phi_L, \mu_0^{\rm in})$ and $q_{,\mu}^- \equiv q_{,\mu}(\phi_s, \mu_0^{\rm in})$ have been defined to simplify notation, while $\Delta H \equiv H^+ - H^-$ and $\Delta J \equiv J^+ - J^-$ are defined by

$$H^+ = \int_0^\infty dx\,\left(c_L - c_0^{\rm in}(x)\right), \quad H^- = \int_{-\infty}^0 dx\,\left(c_0^{\rm in}(x) - c_s\right), \tag{B.79}$$

$$J^+ = \int_0^\infty dx\,\left(q^+ - q(\phi_0^{\rm in}, \mu_0^{\rm in})\right), \quad J^- = \int_{-\infty}^0 dx\,\left(q(\phi_0^{\rm in}, \mu_0^{\rm in}) - q^-\right) \tag{B.80}$$

Cleaning up terms of Eq. (B.78)

It is instructive to simplify the 2nd and 3rd terms on the RHS of Eq. (B.78). To do so, we will neglect at this juncture the q^- and $q_{,\mu}^-$ terms for simplicity, as they are essentially zero for all practical cases. First, use Eq. (B.31) to write $\lim_{\xi\to\pm\infty} c_1^{\rm in}(\xi) = \chi(\phi_{\{s,L\}}, \mu_0^o(0^\pm))\lim_{\xi\to\pm\infty}\mu_1^{\rm in}(\xi)$, and substitute these limits of $\mu_1^{\rm in}(\xi)$ and $c_1^{\rm in}(\xi)$ in terms of their outer fields at the interface (second line of Eq. (B.12) for $\mu_1^{\text{in}}(\xi)$ and a similarly for $c_1^{\text{in}}(\xi)$). Comparing the resulting ξ^0 terms gives

$$c_1^o(0^{\pm}) = \chi(\phi_{\{s,L\}}, \mu_0^o(0^{\pm}))\, \mu_1^o(0^{\pm}) \tag{B.81}$$

Equation (B.81) is used to write Δc_1 in Eq. (B.78) in terms of $\mu_1^o(0^{\pm})$. Also, it is noted that

$$q_{,\mu}^{+} \;=\; \tilde{q}(\phi_L)\chi^L \left\{ \frac{\chi_{,\mu}^L}{\chi^L} \right\} = \frac{\chi_{,\mu}^L}{\chi^L}\, q^{+} \tag{B.82}$$

where $\chi^L \equiv \chi^L(\mu_0^o(0^+))$ and $\chi_{,\mu}^L \equiv \partial\chi(\mu_0^o(0^+))/\partial\mu$. Recalling that $q^+ \partial\mu_0^o(0^+)/\partial\eta = -\bar{v}_0 \Delta c$ (Eq. B.62), and using Eq. (B.81) and Eq. (B.82) gives

$$\bar{v}_0 \Delta c_1 + q_{,\mu}^{+}\, \mu_1^o(0^+) \frac{\partial\mu_0^o(0^+)}{\partial\eta} = \left\{ \chi^L - \frac{\chi_{,\mu}^L}{\chi^L}\, \Delta c \right\} \bar{v}_0\, \mu_1^o(0^+) - \bar{v}_0 \chi_s\, \mu_1^o(0^-) \tag{B.83}$$

Equation (B.83) will be useful for simplifying Eq. (B.78) below.

Final form of the second order flux conservation boundary condition

Using Eq. (B.60) to express $\mu_1^o(0^-)$ in Eq. (B.83) in terms of $\mu_1^o(0^+)$, and substituting the result back into Eq. (B.78) will simplify the expression for the $\mathcal{O}(\epsilon^2)$ correction to the flux conservation condition. Combining this re-expressed form Eq. (B.78) with the $\mathcal{O}(\epsilon)$ correction in Eq. (B.62) gives the final flux conservation condition, accurate to $\mathcal{O}(\epsilon^2)$,

$$\begin{aligned} D_L \chi^L \frac{\partial\mu^o(0^-)}{\partial\eta} &= -\Delta c\, v_n - \left\{ v_0\, \Delta H + D_L \partial_\theta \left(\kappa \partial_\theta \mu_0^o(0^{\pm}) \right) \Delta J \right\} W_\phi \kappa \\ &\quad - \frac{\chi^s\, \Delta F\, W_\phi}{D_L} v_0^2 - \chi^L \left\{ 1 - \frac{\chi^s}{\chi^L} - \frac{\chi_{,\mu}^L}{(\chi^L)^2} \Delta c \right\} \frac{W_\phi}{d_o}\, v_0\, \mu_1^o(0^+) \end{aligned} \tag{B.84}$$

where $\mu^o \approx \mu_0^o + \epsilon\mu_1^o + \mathcal{O}(\epsilon^2)$ and $v_n = v_0 + \epsilon v_1 + \cdots \mathcal{O}(\epsilon^2)$ have been used, and it is noted that Eq. (B.84) has been written back in dimensional units.

Equation (B.84) holds for any binary alloy interface described by the grand potential phase field model derived here. The ΔH, ΔJ and ΔF terms were reported by Almgren [61] and Echebarria and co-workers [10] for ideal binary alloys. The $v_n \Delta c$ term is the lowest order velocity correction to order ϵ. The ΔJ term is generally lower order than $\epsilon = W_\phi/d_o$, meanwhile the effect of both ΔJ and ΔH vanishes for flat interfaces. The ΔF term can be factored into the $v_n \Delta c$ term (to lowest order) to adjust the concentration jump to account for a chemical potential jump at the interface, which is relevant to rapid solidification. The last term in Eq. (B.84) is new, it is identically zero only for ideal binary alloys, but not in general. It can be dropped here as it is of order $\mathcal{O}(\epsilon)$.

Bibliography

[1] J. B. Collins and H. Levine. *Phys. Rev. B*, 31:6119, 1985.

[2] P. C. Hohenberg and B. I. Halperin. *Rev. Mod. Phys.*, 49:435, 1977.

[3] A. Karma and W. J. Rappel. Quantitative phase-field modeling of dendritic growth in two and three dimensions. *Phys. Rev. E*, 57:4323–4349, 1998.

[4] J. S. Langer. *Rev. Mod. Phys.*, 52:1, 1980.

[5] G. Caginalp and E. Socolovsky. *SIAM J. Sci. Comp.*, 15:106, 1991.

[6] N. Provatas, N. Goldenfeld, and J. Dantzig. Efficient computation of dendritic microstructures using adaptive mesh refinement. *Phys. Rev. Lett.*, 80:3308–3311, 1998.

[7] N. Provatas, J. Dantzig, and N. Goldenfeld. *J. Comp. Phys.*, 148:265, 1999.

[8] J. A. Warren and W. J. Boettinger. *Acta Metall. Mater. A*, 43:689, 1995.

[9] A. Karma. *Phys. Rev. Lett*, 87:115701, 2001.

[10] B. Echebarria, R. Folch, A. Karma, and M. Plapp. *Phys. Rev. E.*, 70:061604–1, 2004.

[11] J. C. Ramirez, C. Beckermann, A. Karma, and H. J. Diepers. *Phys. Rev. E*, 69:051607, 2004.

[12] C. Tong, M. Greenwood, and N. Provatas. *Phys. Rev. E*, 77:1, 2008.

[13] Nana Ofori-Opoku and Nikolas Provatas. A quantitative multi-phase field model of polycrystalline alloy solidification. *Acta Materialia*, 58(6):2155–2164, 2010.

[14] A. G. Khachaturyan. *Theory of Structural Transformations in Solids.* Wiley-Interscience Publications (New York), 1983.

[15] K. R. Elder, Zhi-Feng Huang, and Nikolas Provatas. Amplitude expansion of the binary phase-field-crystal model. *Phys. Rev. E*, 81:011602, Jan 2010.

[16] Zhi-Feng Huang, K. R. Elder, and Nikolas Provatas. Phase-field-crystal dynamics for binary systems: Derivation from dynamical density functional theory, amplitude equation formalism, and applications to alloy heterostructures. *Phys. Rev. E*, 82:021605, Aug 2010.

[17] Kuo-An Wu, Ari Adland, and Alain Karma. Phase-field-crystal model for fcc ordering. *Phys. Rev. E*, 81:061601, 2010.

[18] Nana Ofori-Opoku, Jonathan Stolle, Zhi-Feng Huang, and Nikolas Provatas. Complex order parameter phase-field models derived from structural phase-field-crystal models. *Phys. Rev. B*, 88:104106, Sep 2013.

[19] D. Fan, S. P. Chen, L.-Q. Chen, and P. W. Voorhees. *Acta Materialia*, 50:1897, 2002.

[20] J. Z. Zhu, T. Wang, A. J. Ardell, S. H. Zhou, Z. K. Lui, and L. Q. Chen. *Acta Materialia*, 52:2837, 2004.

[21] J. Zhu, T. Wang, S. Zhou, Z. Liu, and L.-Q. Chen. *Acta Materialia*, 52:833, 2004.

[22] J. Tiaden, B. Nestler, H. Diepers, and I. Steinbach. *Physica D*, 115:73, 1998.

[23] S. G. Kim, W. T. Kim, and T. Suzuki. *Phys. Rev. E*, 60:7186, 1999.

[24] S. G. Kim. *Acta Materialia*, 55:4391, 2007.

[25] L. Q. Chen and W. Yang. *Phy. Rev. B*, 50:15752, 1994.

[26] A. Kazaryan, Y. Wang, S. A. Dregia, and B. R. Patton. *Phys. Rev. B*, 61:14275, 2000.

[27] C. Shen, Q. Chen, Y. H. Wen, J. P. Simmons, and Y. Wang. *Scripta Materialia*, 50(7):1023–1028, 2004.

[28] N. Moelans, B. Blanpain, and P. Wollants. *Phys Rev Lett*, 101(025502), 2008.

[29] N. Moelans, B. Blanpain, and P. Wollants. *Phy. Rev. B*, 78:024113, 2008.

[30] B. Morin, K. R. Elder, M. Sutton, and M. Grant. *Phys. Rev. Lett.*, 75:2156, 1995.

[31] R. Kobayashi, J. A. Warren, and W. C. Carter. *Physica D*, 119:415, 1998.

[32] R. Kobayashi, J. A. Warren, and W. C. Carter. *Physica D.*, 140:141, 2000.

[33] R. Kobayashi and J. A. Warren. *Physica A*, 356:127, 2005.

[34] L. Gránásy, T. Börzsönyi, and T. Pusztai. Crystal nucleation and growth in binary phase-field theory. *Journal of Crystal Growth*, 237(1813–1817), 2002.

[35] Badrinarayan P. Athreya, Nigel Goldenfeld, Jonathan A. Dantzig, Michael Greenwood, and Nikolas Provatas. Adaptive mesh computation of polycrystalline pattern formation using a renormalization-group reduction of the phase-field crystal model. *Phys. Rev. E*, 76:056706, Nov 2007.

[36] Yongmei M. Jin and Armen G. Khachaturyan. Atomic density function theory and modeling of microstructure evolution at the atomic scale. *Journal of Applied Physics*, 100(1):013519, 2006.

[37] K. R. Elder, N. Provatas, J. Berry, P. Stefanovic, and M. Grant. *Phys. Rev. B.*, 75:064107, 2007.

[38] Nikolas Provatas and Ken Elder. *Phase-Field Methods in Materials Science and Engineering*. Wiley-VCH Verlag GmbH & Co. KGaA, 2010.

[39] Sami Majaniemi, Nana Ofori-Opoku, and Nikolas Provatas. 2017.

[40] Robert Spatschek and Alain Karma. Amplitude equations for polycrystalline materials with interaction between composition and stress. *Phys. Rev. B*, 81:214201, June 2010.

[41] I. Steinbach, F. Pezzolla, B. Nestler, M. Seebelberg, R. Prieler, and G. J. Schmitz. *Physica D*, 94:135, 1999.

[42] H. Garcke, B. Nestler, and B. Stoth. *SIAM J. Appl. Math*, 60:295, 1999.

[43] B. Nestler, H. Garcke, and B. Stinner. *Phys. Rev. E*, 71:041609–1, 2005.

[44] B. Bottger and I. Steinbach. *Acta Materialia*, 54:2697, 2006.

[45] I. Steinbach. *Apel M. Physica D*, 217:153, 2006.

[46] J. Eiken, B. Böttger, and I. Steinbach. *Phys Rev E*, 73:066122, 2006.

[47] I. Steinbach. *Modelling Simul. Mater. Sci. Eng.*, 17:073001, 2009.

[48] Nele Moelans. *Acta Materialia*, 59:1077, 2011.

[49] R. Folch and M. Plapp. *Phys. Rev. E.*, 72:011602, 2005.

[50] A. Choudhury and B. Nestler. *Phys. Rev. E*, 85:021602, 2012.

[51] Mathis Plapp. Unified derivation of phase-field models for alloy solidification from a grand-potential functional. *Phys. Rev. E*, 84:031601, 2011.

[52] Johannes Hotzer, Marcus Jainta, Philipp Steinmetz, Britta Nestler, Anne Dennstedt, Amber Genau, Martin Bauer, Harald Kostlerc, and Ulrich Rudec. Large scale phase-field simulations of directional ternary eutectic solidification. *Acta Materialia*, 93:194, 2015.

[53] M. J. Aziz and T. Kaplan. *Acta Metall.*, 36:2335, 1988.

[54] M. J. Aziz and W. J. Boettinger. *Acta Metall. Mater.*, 42:527, 1994.

[55] N. A. Ahmad, A. A. Wheeler, W. J. Boettinger, and G. B. McFadden. *Phys Rev E*, 58:3436, 1998.

[56] Martin Grant. Dirty tricks for statistical mechanics: time dependent things. Lecture Notes for Advanced Statistical Physics, Version 0.8, August 2005.

[57] Jan-Olof Andersson and John Agren. Models for numerical treatment of multicomponent diffusion in simple phases. *J. Appl. Phys.*, 72(4):1350, 1992.

[58] Jonathan A. Dantzig and Michel Rappaz. *Solidification*. EPFL Press, 2009.

[59] G. Caginalp and P. C. Fife. *Phys. Rev. B*, 11:7792, 1986.

[60] G. Caginalp. *Phys. Rev. A*, 39:5887, 1989.

[61] R. Almgren. *SIAM J. Appl. Math.*, 59:2086, 1999.

[62] Kuo-An Wu and Alain Karma. Phase-field crystal modeling of equilibrium bcc-liquid interfaces. *Phys. Rev. B*, 76:184107, Nov 2007.

[63] Sami Majaniemi and Nikolas Provatas. Deriving surface-energy anisotropy for phenomenological phase-field models of solidification. *Phys. Rev. E*, 79(1):011607, 2009.

[64] Nikolas Provatas and Sami Majaniemi. Phase-field-crystal calculation of crystal-melt surface tension in binary alloys. *Phys. Rev. E*, 82(4):041601, 2010.

[65] Hao Zhang, David J. Srolovitz, Jack F. Douglas, and James A. Warren. *PNAS*, 106(19):7735, 2009.

[66] Laishan Yang, Zhibo Dang, Lei Wang, and Nikolas Provatas. Improvements to multi-order parameter and multi-component phase models of polycrystalline solidification. In press (2021), DOI 10.1016/j.jmst.2021.06.017 .

[67] Q. Bronchart, Y. Le Bouar, and A. Finel. *Phys Rev Lett*, 100:015702, 2008.

[68] Chakin and Lubensky. *Principles of Condensed Matter Physics*, page 629. North Holland, Amsterdam, 1987.

[69] Blas Echebarria, Alain Karma, and Sebastian Gurevich. *Phys. Rev. E*, 81:021608, 2010.

[70] L. Landau and E. Lifsitz. *Statistical Physics*, 3rd ed. Butterworth-Heinemann, Oxford,, 1987.

[71] J. Bragard, A. Karma, Y. H. Lee, and M. Plapp. *Interface Sci.*, 10:121, 2002.

[72] J. Heulens, B. Blanpain, and N. Moelans. *Acta Materialia*, 59:2156, 2011.

[73] H. L. de Villiers Lovelock. Powder/processing/structure relationships in wc-co thermal spray coatings: a review of the published literature. *Journal of Thermal Spray Technology*, 7(3):357–373, 1998.

[74] Pierre Fauchais, Ghislain Montavon, and Ghislaine Bertrand. From powders to thermally sprayed coatings. *Journal of Thermal Spray Technology*, 19(1–2):56–80, 2010.

[75] Enrique J. Lavernia and Tirumalai S. Srivatsan. The rapid solidification processing of materials: science, principles, technology, advances, and applications. *Journal of Materials Science*, 45(2):287–325, 2010.

[76] S. Y. Hu, J. Murray, H. Weiland, Z. K. Liu, and L. Q. Chen. Thermodynamic description and growth kinetics of stoichiometric precipitates in the phase-field approach. *Calphad*, 31(2):303–312, 2007.

[77] Martin Walbrühl. Diffusion in the liquid co-binder of cemented carbides: Ab initio molecular dynamics and dictra simulations, 2014.

[78] Zengyun Jian, Kazuhiko Kuribayashi, and Wanqi Jie. Solid-liquid interface energy of metals at melting point and undercooled state. *Materials Transactions*, 43(4):721–726, 2002.

[79] Mingheng Li, Dan Shi, and Panagiotis D Christofides. Modeling and control of HVOF thermal spray processing of wc–co coatings. *Powder Technology*, 156(2-3):177–194, 2005.

[80] Tatu Pinomaa. Thermal spray process modeling: A focus on gas dynamics and particles in-flight. Master's Thesis, Aalto University, 2013.

[81] R. C. Brower, D. Kessler, J. Koplik, and H. Levine. *Phys. Rev. Lett.*, 51:1111, 1983.

[82] E. Ben-Jacob, N. Goldenfeld, J.S. Langer, and G. Schön. *Phys. Rev. Lett.*, 51:1930, 1983.

[83] E. Ben-Jacob, N. Goldenfeld, B. G. Kotliar, and J. S. Langer. *Phys. Rev. Lett.*, 53:2110, 1984.

[84] D. Kessler J. Koplik and H. Levine. *Phys. Rev. A*, 30:3161, 1984.

[85] D. A. Kessler and H. Levine. *Phys. Rev. A.*, 39:3041, 1989.

[86] E. Brener and V. I. Melnikov. *Adv. Phys.*, 40:53, 1991.

[87] Y. Pomeau and M. Ben Amar. *Solids Far from Equilibrium*, page 365. Edited by C. Godreche (Cambridge Press), 1991.

[88] S. L. Sobolev. *Phys. Rev. A*, 199:383–386, 1995.

[89] S. L. Sobolev. *Phys. Rev. E*, 55(6), 1997.

[90] A. A. Wheeler, W. J. Boettinger, and G. B. McFadden. *Phys. Rev. E*, 47:1893, 1993.

[91] W. J. Boettinger, A. A. Wheeler, B. T. Murray, and G. B. McFadden. *Materials Science and Engineering: A*, 178:217.

[92] D. Turnbull. *J. Phys. Chem.*, 66:609, 1962.

[93] J. W. Cahn. *Acta. Metall.*, 10:789–798, 1962.

[94] J. C. Baker and J. W. Cahn. *Solidification.* ASM, Metals Park, OH, 1971.

[95] Peter Galenko. *Phy. Rev. B*, 65:144103, 2002.

[96] Tatu Pinomaa and Nikolas Provatas. Quantitative phase field modeling of solute trapping and continuous growth kinetics in quasi-rapid solidification. *Acta Materialia*, 168:167–177, 2019.

[97] Tatu Pinomaa, Joseph T McKeown, Jörg MK Wiezorek, Nikolas Provatas, Anssi Laukkanen, and Tomi Suhonen. Phase field modeling of rapid resolidification of al-cu thin films. *Journal of Crystal Growth*, 532:125418, 2020.

Index

Note: Page numbers in *italics* refer to figures.